Math Workout for the New GRE®

2nd Edition

The Princeton Review®

Math Workout for the New GRE®

2nd Edition

Revised and Updated for the New GRE

Doug French

PrincetonReview.com

Random House, Inc. New York

The Princeton Review, Inc.
111 Speen Street, Suite 550
Framingham, MA 01701
E-mail: editorialsupport@review.com

ISBN 978-0-375-42820-3
ISSN 1558-9692

Editor: Heather Brady
Production Editor: Michael Breslosky
Production Coordinator: Deborah A. Silvestrini
Illustrations: The production department of The Princeton Review

Printed in the United States of America on partially recycled paper.

10 9 8 7 6 5 4 3 2 1

Editorial

Rob Franek, VP Test Prep Books, Publisher
Seamus Mullarkey, Associate Publisher
Laura Braswell, Senior Editor
Selena Coppock, Editor
Heather Brady, Editor

Random House Publishing Team

Tom Russell, Publisher
Nicole Benhabib, Publishing Manager
Ellen L. Reed, Production Manager
Alison Stoltzfus, Associate Managing Editor

Acknowledgments

I'm indebted to Ellen Mendlow for her patient leadership (and keen eye for writing talent) and to Tricia McCloskey, the R&D guru who was a rare source of answers in a maelstrom of questions.

And last, thanks to my family, who understood when Daddy had to run off and write.

The Princeton Review would also like to thank the following people for their invaluable expertise in revising and updating the second edition of this book:

Elizabeth Aamot, Jennifer Amerkhanov, Michael Chapian, Vanessa Coggshall, Loren Dunn, Peter Hanink, Karen Hoover, Alison Howard, Michael Levy, Lisa Mayo, Jerome J. O'Neill, Elizabeth Owens, Debbi Reynolds, Tim Ricchuiti, Lisa Rothstein, David Stoll, Scott Thompson, and particularly Curtis Retherford, Neill Seltzer, Jennifer Downey, and Liz Rutzel.

Contents

1 Introduction ... 1

2 The End of Mathophobia ... 17

3 The Nuts and Bolts ... 29

4 Algebra, and How to Get Rid of It ... 85

5 Charts and Crafts ... 155

6 Math in the Real World ... 181

7 Geometry ... 211

8 The Rest of the Story ... 249

9 Sample Section 1 ... 287

10 Answers and Explanations to Sample Section 1 ... 293

11 Sample Section 2 ... 299

12 Answers and Explanations to Sample Section 2 ... 305

13 Glossary of Math Terms ... 311

About the Author ... 321

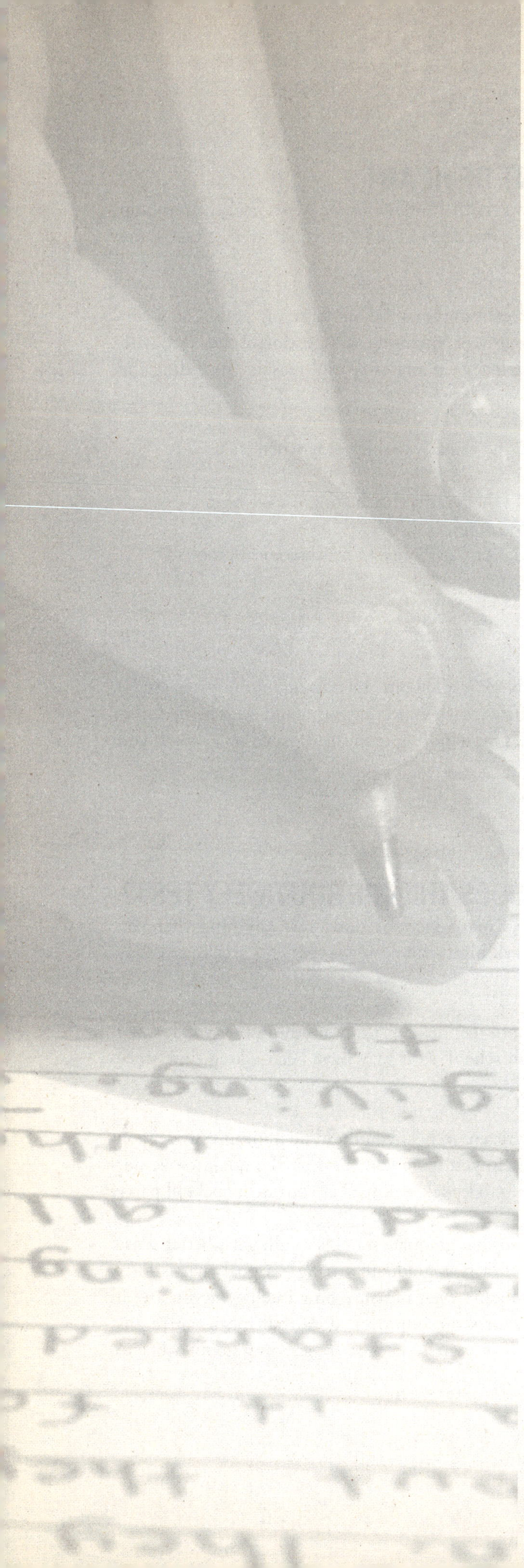

Chapter 1
Introduction

ADVICE FOR THE FAINT OF HEART

Welcome to The Princeton Review's *Math Workout for the New GRE*, the one-stop shop for all of the mathematical knowledge and practice you'll need to tackle the Math section of the GRE effectively.

You've bought this book, which means you may be one of many grad-school candidates who are approaching the math, or "quantitative," portion of the GRE with a little bit of trepidation. This might be for any of several reasons, including the following:

- You come in contact with the word "variable" only when it's used to describe the weather.
- Your first thought about Pythagoras is that he might have been a character in *The Lord of the Rings*.
- You regard "standard deviation" as more of a psychological problem than a mathematical one.

If any of the above pertain to you, you're definitely not alone.

But don't worry, that's what this book is all about. Its two main objectives are (1) to give you an overview of all of the math concepts you could see on the GRE, and (2) to give you simple strategies for handling even the most complex math you could encounter on test day.

WHAT KIND OF MATH DOES THE GRE ROUTINELY TEST?

The good news is that the GRE's math sections don't test anything that you learned after your sophomore year of high school, so the concepts aren't extremely advanced.

The bad news is that the GRE's math sections don't test anything that you learned after your sophomore year of high school, so it may have been a long time since you studied them.

That's largely why this book was written: to help you build up an impressive canon of math knowledge that will (1) help you score your best on the quantitative portion of the GRE, and (2) set you up to knock 'em dead at your next cocktail party.

The GRE supposedly was written so that graduate schools might get a better sense of an applicant's ability to work in a post-graduate setting—a goal that is lofty and unrealistic at best. The test doesn't even measure how intelligent you are; if you take a test-prep course and your score improves, does that mean you're any smarter? Nope. Yet you can improve your score just by learning about what to expect on the GRE.

All the GRE really tests is how well you take the GRE.

Succeeding on the quantitative portion of the GRE—or any standardized test, for that matter—is as much about relearning math concepts as it is about modifying the way you think. There are several very important skills to cultivate when you're preparing to take the GRE, and each of them is attainable with the right guidance, a strong work ethic, and a healthy dose of optimism.

We'll discuss the math basics you'll need for the GRE, but if you need a quick reference, consult the glossary at the back of the book.

The Layout of the Test

Let's talk about the different sections of the GRE. The GRE contains five scored sections:

- One 60-minute Analytical Writing section, which contains two essay questions
- Two 30-minute Verbal sections, which contain approximately 20 questions each
- Two 30-minute Math sections, which contain approximately 20 questions each

The first section will always be the Analytical Writing section, followed by the Math and Verbal sections, which can appear in any order. All of the Verbal questions are multiple choice. The Math questions are mostly multiple choice with some numeric entry questions, which require typing in an answer.

You will be able to see your Verbal and Math scores immediately upon completion of the test, but you will have to wait about two weeks before your Analytical Writing section is scored.

Scores are given on a scale from 130 to 170, in 1-point increments. The questions within each section are always worth the same amount of points. So the easy questions in a section are just as important to get right as the hard questions in a section.

Once you've completed one Math section, the GRE will use your score on that section to determine which questions to give you in the next Math section. The same applies for the two Verbal sections. (This doesn't really affect how you will approach the test, so don't worry about it too much.)

You will get a 1-minute break—enough time to close your eyes and catch a breath—between each section. You will also get a full 10-minute break after the third section. Be sure to use it to visit the bathroom, take a drink of water, refresh your mind, and get ready for the rest of the exam.

Experimental Section

Here's where ETS, the makers of the test, start to get mean. In addition to the five scored sections listed above (one Analytical Writing, two Math, two Verbal), you may also have an unscored experimental section. This section is almost always a Math or Verbal section. It will look exactly like the other Math or Verbal sections, but it won't count at all toward your score. ETS administers the experimental section to gather data on questions before they appear on real GREs.

Thus, after your Analytical Writing section you will probably see five—not four—multiple-choice sections: either three Verbal and two Math, or two Verbal and three Math, depending on whether you get a Verbal or Math experimental section. These sections can come in any order. You will have no way of knowing which section is experimental, so you need to do your best on all of them. Don't waste time worrying about which sections count and which section does not.

Here is how a typical GRE might look:

Analytical Writing – 60 minutes
Verbal – 30 minutes
10-minute break
Math – 30 minutes
Math – 30 minutes
Verbal – 30 minutes
Math – 30 minutes

Remember, the Analytical Writing section will always be first, and it will never be experimental. In the above example, the two Verbal sections will be scored, but out of the three Math sections only two will be scored. One of the three is an experimental section, but we don't know which one. Of course, on your GRE you might see three Verbal sections instead, meaning one of your Verbal sections is experimental, and they may come in any order. Be flexible, and you'll be ready for the test no matter what order the sections come in. In fact, the test makers may not even include an experimental section! If so, count your lucky stars that you didn't have to waste your time on a meaningless section.

Research Section

At the end of the test, you may also have an unscored Research section. At the beginning of this section, you will be told that it is an unscored Research section, used only to help develop and test questions for the GRE. If you want to skip it, you have the option of skipping it. They normally offer some sort of prize to induce people to take it, but by that point in the test you will probably be exhausted. If they offer you a Research section, just go ahead and decline, get your scores, and go home.

MATH OVERVIEW

There are three main skills that we will emphasize throughout this book: *keep your hand moving, take the easy test first,* and *be prepared to walk away.* These are not necessarily what you would naturally do while taking a test, so you'll have to force yourself to apply these skills as you work through the problems in this book and as you take practice tests. If you do, you'll find that once you get to the real test your body and brain already know how to tackle each question, and you'll be able to breathe a bit easier.

Most people assume that the Math section is about thinking. It's not. It's about *doing.* When you get stuck, resist the urge to sit and stare. Instead, get your hand moving.

Keep Your Hand Moving

You'll get about six pieces of scratch paper to use for the test. Use them. Use all of them, and then get more. Don't just use scratch paper for multiplying or doing long division. Use your scratch paper for every single part of the problem, from beginning to end.

You will not solve the problems in your head, and you will not solve them on the screen. You will solve them on the scratch paper. As soon as you click Next, and see a question, start writing.

We'll go into exactly what to write for each question later in the book. For now, you should get into the habit of writing the following:

- **Question Number:** You may need to leave a question and come back to it. If so, you may have some calculations that you can reuse. If you've got the question number written down, you can easily see what work goes with which question.
- **Answer Choices:** For most questions, this will mean simply writing A B C D E vertically on the left side of your scratch paper. As you work each question, you may realize that certain answer choices are definitely wrong. In that case, cross them off on your scratch paper.
- **Problem Set Up:** On the left side of your paper, next to the answer choices, write down information as you read the problem. If it's a geometry question, redraw the figure. If the question says that Bob has 142 oranges in his grove, and Sue has 219 oranges in her grove, then immediately write down "Bob = 142 oranges" and "Sue = 219 oranges." Don't keep anything in your head.
- **Calculations:** As you work through the problem, you will probably need to the use the onscreen calculator. As you do so, write down every single calculation on your piece of paper, off to the right side. Don't enter in $216 \times 3 \div 4$ all at once. First, do 216×3 on the calculator, and write down the result, 648. Then, do $648 \div 4$, and write down the result, 162. It's easy to make small mistakes with the calculator if you're not careful.

- **A Horizontal Line:** After each problem, draw a line to separate it from the next problem. This will keep your work organized, and prevent you from accidentally using numbers or information for one problem while solving another.

Your scratch paper may end up looking something like this:

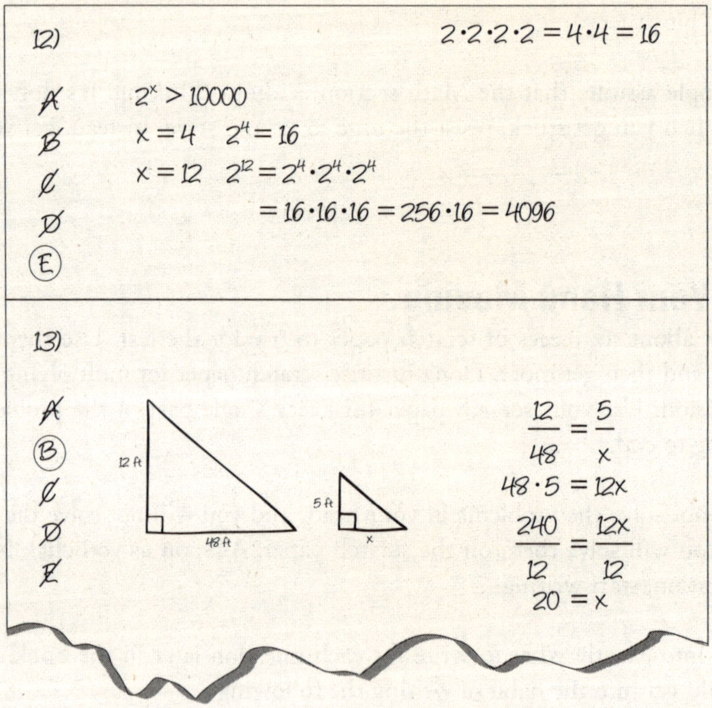

Take the Easy Test First

All questions within a given section are worth the same amount of points. Many people rush through the easy questions so they can spend more time on the hard questions. However, if easy questions are worth just as much as hard questions, why not focus just as much on them?

There are a certain number of questions on the GRE that you can easily answer correctly. As soon as you read through them, you know what they're asking and how to get to the answer. Your job is to answer all of those questions first. Don't rush through them, because you can't afford to get these questions wrong. These are practically free points, as long as you're paying attention. Use your scratch paper and read carefully.

Save the hard questions for later. You can always return to them, even if it's just for a last-second guess. The goal with your first pass through any section is to get as many points as you can, without any mistakes. Once you've done that, you can use the time remaining to return to the other, harder questions. You'll find that after a second look, some of the hard questions are easier than you initially thought. Go ahead and do those questions now. Some of the questions you thought were going to be hard are, in fact, hard questions. Leave those. You'll come back with any time remaining and either work through them or eliminate answers and guess.

> Easy questions are worth the same as hard questions. Work easy questions carefully, so you don't get any wrong.

Be Prepared to Walk Away

At the top of the screen are buttons labeled Mark, Review, and Next. Any question you're not sure about, click Mark, then click Next and move on. If you click on Review, you'll see a screen like this:

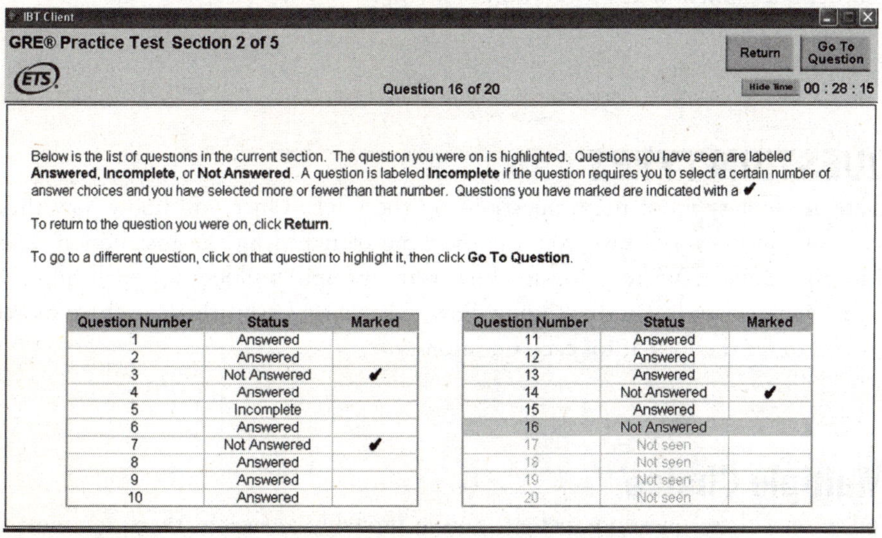

Here you can see every question you haven't answered, and every question you marked to come back to later. If you need to return to any question, you can click on that question on the review screen and you'll be brought right to it.

Why is all this so important? Because you can always move on. If you read a question and you don't immediately know what to do, move on. If a question seems particularly difficult, move on. If you start working through a question and realize you aren't getting any closer to the answer, move on. If you work through a question and the answer you got isn't among the answer choices, move on.

Your first pass is all of the easy questions. Your second pass is the harder questions, and these are the questions you're most likely to misread. Once you've read a question one way, it's hard to get your brain to read that question any other way. So if you're not sure what the question is asking, if you realize you're doing a lot more math than you normally do for GRE questions, or if you get an answer that isn't one of the answer choices, then move on. Do a couple other questions, give your brain a chance to shift gears, and then come back to it. Don't look back at your old scratch paper yet. Reread the question and take notes on your scratch paper as if it's the first time you've seen this question.

Any time you stall while working on a question, move on. Do not let yourself get stuck. Sitting and staring blankly at the computer screen does not help. Move to another question, and keep your hand moving.

QUESTION TYPES

There are four types of math questions on the GRE. Once you know how these questions work, you'll save yourself the time of rereading the instructions each time they appear. We're going to show you a sample problem for each question type. Don't worry if you don't know how to solve these yet; these are here mostly for you to see the format for each question type.

Multiple Choice

You've seen these questions before. You've probably answered them for most of your life. Multiple-choice questions are questions which have five answer choices. You have to select one answer choice and then click Next.

The answer bubbles for these questions will always be round. Whenever a question has circular bubbles, you must select one and only one answer and then click Next to continue.

Get used to thinking of each answer choice as (A), (B), (C), (D), or (E). As soon as you see a multiple-choice question, write down A B C D E vertically on the left side of your scratch paper.

Question 1 of 4

If c is the largest prime number smaller than 22, and d is the smallest prime number larger than 35, then $c + d =$

- ○ 33
- ○ 41
- ○ 50
- ○ 56
- ○ 58

Here's How to Crack It

Since this is a multiple-choice question, write down the question number and A B C D E on your scratch paper. The question says that *c is the largest prime number smaller than 22.* Check each number smaller than 22, starting with 21. Is 21 prime? Nope. Neither is 20. 19 is prime, so 19 is the largest prime number that is still smaller than 22. Write down $c = 19$. Now try to find d. It must be larger than 35, and prime, so it can't be 36. The next number, 37, is prime, because the only numbers we can divide 37 by are 37 and 1, so write down $d = 37$. The question asks for $c + d$, which is $19 + 37 = 56$, answer (D).

Quantitative Comparison

We'll normally call these questions Quant Comp for short. These questions are variations of the basic multiple-choice question. You will be given Quantity A in one column and Quantity B in another column, and you must select one of four answers:

- ○ Quantity A is greater.
- ○ Quantity B is greater.
- ○ The two quantities are equal.
- ○ The relationship cannot be determined from the information given.

These answer choices are the same for every single Quantitative Comparison question. (A) means that Quantity A is always, always greater, (B) means that Quantity B is always greater, (C) means that the two quantities are always equal, and (D) means that we're not sure: Sometimes Quantity A is bigger, sometimes Quantity B is bigger, or sometimes they're the same.

Since these questions have round answer bubbles, you'll select only one answer: (A), (B), (C), or (D). As soon as you see one of these questions, write down A B C D vertically on the left side of your scratch paper.

Before we do a sample question, there's one important thing you should know about answer choice (D): It can never be the answer for straight calculation questions. For instance, if Quantity A is 10^{12} and Quantity B is 5^{24}, then a calculator or computer could solve that question, right? Whatever the answer is, it will always be one particular number for Quantity A, and one particular number for Quantity B. Sure, those numbers are super hard to find with the calculator, and we would have to use some clever tricks to actually find the answer, but the answer can't be (D). If there are no variables, and the problem is simply about doing calculations, the answer can't be (D) for a Quant Comp question, because the relationship *can* be determined, it may be a pain to determine it.

Question 2 of 4

Square *ABCD* contains point *E*. The area of *ABCD* is 100.

Quantity A	**Quantity B**
The length of line *DE*	15

○ Quantity A is greater.

○ Quantity B is greater.

○ The two quantities are equal.

○ The relationship cannot be determined from the information given.

Here's How to Crack It

Since this is a Quant Comp question, write down A B C D vertically on the left side of your paper. Draw square *ABCD*. The area of *ABCD* is 100, which means each side must be 10. Label the sides on your figure. The problem states that point *E* is somewhere within the square, but it doesn't say exactly where. It could, for instance, be really close to point *D*, only 1 unit away. In that case, *DE* would definitely be smaller than 15, so cross off answers (A) and (C) on your scratch paper: Quantity A isn't always greater, and the two quantities aren't always equal. Now we have to find out if we can make Quantity A greater than Quantity B. What if we put point *E* all the way on the other side of the square? Go ahead and redraw the square, but now put *E* next to point *B*. How far away are *E* and *D* now? Well, since it's a square, the distance from *E* to *D* makes a 45-45-90 triangle, which means that the distance is less than $10\sqrt{2} \approx 10 \times 1.4 = 14$. In that case, *DE* is still less than 15, so the answer is (B). No matter where we put point *E*, it is always less than 15 units away from point *D*. Your scratch paper should look something like this:

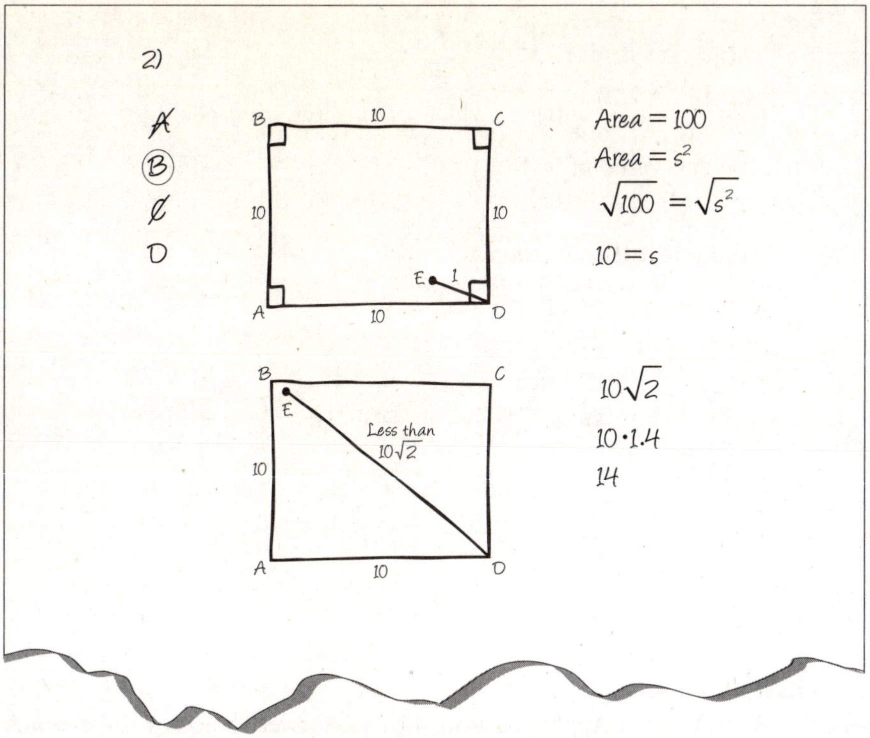

All That Apply

These are multiple-choice questions which will have any number of answer choices (generally three or five, but sometimes more), and you will have to select all the answers that apply to the question. The answer choices for these questions are always squares. The question will typically state to select *all* values or statements that apply.

Note that there's no partial credit for these questions. You must choose every single answer that works, or you get no credit for that question. There will always be at least one answer for these questions, but there may be only one answer that works. Or two. Or three. Or all eight answer choices could work.

When you see these questions, write down letters (A, B, etc.) for each answer choice. As you work through the problem, put a check mark next to each answer choice that works, and cross off any answer choice that doesn't work. Before you click Next, double check that you have selected all the answers you have written on your scratch paper, and none of the answers that you crossed off.

Question 3 of 4

If $\frac{12^{13}}{x}$ is an integer, which of the following could be the value of x ?

*Indicate **all** such values.*

☐ 8

☐ 10

☐ 36

☐ 64

☐ 112

☐ 432

Here's How to Crack It

Since this is an All That Apply question with six answer choices, write down A B C D E F vertically on the left side of your paper. Notice that they're asking about 12^{13}? That's way too big to enter into our calculator, so we'll have to do something other than just brute force calculation. For the fraction to be an integer, x will have to be a factor of 12^{13}. For instance, if x were 12, then the fraction would be an integer because the 12 in the denominator would cancel out with one of the 12s in the numerator. x could also be 2, because 2 is one of the prime factors of 12.

Let's break 12 down into prime factors. Once we know the prime factors of 12^{13}, then we'll know that any answer choice that has some of those same prime factors is definitely a factor of 12^{13}. The prime factors of 12 are $2 \times 2 \times 3 = 2^2 \times 3$. Therefore the prime factors of 12^{13} are $(2^2 \times 3)^{13} = 2^{26} \times 3^{13}$. So any number with up to 26 2s as prime factors will fit into 12^{13}, as will any number with up to 13 3s as prime factors.

(A) is 8. The prime factors of 8 are $2 \times 2 \times 2$. Since 12^{13} definitely contains 2^3, put a check next to (A). (B) is 10, which is the same as 2×5. Although that 2 will fit into 12^{13}, 5 will not. Since 12^{13} is only made up of 2s and 3s, that 5 has nothing to cancel out with. Cross off answer (B), because x cannot be 10; if it were, the fraction would not be an integer. (C) is 36, which has prime factors of $2^2 \times 3^3$, so put a check mark next to it. (D) is 64, which is 8×8, which is $2^3 \times 2^3$, which is 2^6. Put a check mark next to it. (E) is 112. Since 112 is even, start by dividing by 2: $112 = 2 \times 56 = 2 \times 8 \times 7 = 2 \times 2^3 \times 7 = 2^4 \times 7$. Although the 2^4 will divide evenly into 12^{13}, that 7 will not, because 7 is not a prime factor of 12^{13}. Cross off (E). The only one left is (F), which is 432. Break down 432 to its prime factors and you'll get $2^4 \times 3^3$, which definitely fits into 12^{13}. The answers are therefore (A), (C), (D), and (F).

Numeric Entry

These questions don't give you any answer choices at all. Instead, you'll be given a question and an empty box to type a number in. Your answer could be an integer, a decimal, positive, or negative. Never round your answer unless the question asks you to, or it's a question that cannot have decimal answers (the number of children on a school bus, for instance).

The GRE will give you the correct units—they'll be right there next to the box. So before you submit your answer, be sure that it uses the proper units. Be extra careful if problems involve dollars and cents, ounces and pounds, feet and inches, percents, and other common increments that have sub-increments.

Question 4 of 4

Jamal paid $236.30 to rent a car last month. The total consisted of a flat fee of $95 and a charge of $0.075 for every tenth of a mile. For how many miles was Jamal charged?

[ ] miles

Click on the answer box, then type in a number.
Backspace to erase.

Here's How to Crack It

Start at the beginning: If Jamal paid a total of $236.30 and $95 of that was the flat fee, then the cost of the mileage must have been $236.30 – $95.00, or $141.30. Now, here's the crucial conversion: If the rental company charged $0.075 per tenth of a mile, then each mile cost $0.075 × 10, or $0.75. Divide $141.30 by $0.75, and your work is done. The answer is 188.4 miles.

If the answer will be in the form of a fraction, you will see two boxes, one on top of the other, like this:

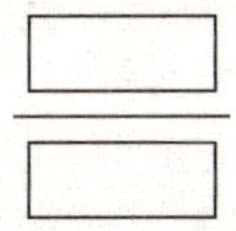

Each of these boxes can hold a maximum of five characters, so your fraction can get pretty complex. For example, if you solve a problem and the answer is $\frac{123}{347}$, you should enter the numerator and denominator separately, like this:

$$\frac{\boxed{123}}{\boxed{347}}$$

HOW TO USE THIS BOOK

In *Math Workout for the New GRE*, we focus solely on the math portions of the GRE. This book includes more than two hundred sample questions (including two sample GRE quantitative sections, complete with answers and explanations) on which to practice all of the new techniques you learn. This book is called a "workout" because if you resolve to "feel the burn" of diligent mental exertion, you won't just memorize a bunch of new techniques. Instead, you'll absorb them into your subconscious so completely that you will use them automatically.

Trust the Techniques

As we'll discuss in Chapter 2, you're about to do a lot of work toward *changing the way you think about taking this test.* To do that, you should be prepared to let go of a number of presumptions and give yourself over to the techniques, which we've designed to conserve your thinking power and greatly reduce the chance that you'll make careless errors come test day. Some of the techniques might seem a little strange or counterintuitive at first, but trust us: Part of the secret to a better score on a standardized test is to think in a non-standardized way.

When we encounter stress, we are hard-wired to fall back on our instincts to protect ourselves. If you start to feel anxious as you take the GRE, you might be tempted to abandon the new techniques in favor of whatever methods you used in order to get through high school and college—methods that won't be as useful.

So when you work with practice questions, be sure to practice using our techniques over and over again. Once you see them working, you'll build enough faith in them to let them *replace* your old habits. Soon you'll summon them without thinking.

Set Up a Schedule—and Stick to It

When you've registered to take the GRE, it's important to keep preparing for the test almost every day. Cramming for eight hours on a Sunday and then leaving the book alone for a week won't be very useful because, like anything else, your new skills will atrophy with disuse. It will be far more effective if you set aside one hour per day to study. You might also want to consult our online test companion at **PrincetonReview.com**; this will help you set tangible goals and plan a reasonable schedule to achieve them.

When you set up your work regimen, keep these things in mind.

- As you work, look for patterns in the types of questions that you frequently answer correctly and patterns in the types you keep getting wrong. This will help you pinpoint your strengths and weaknesses and guide you to the areas in which you need more practice.
- Again, be sure to use the new techniques. If you read up about all these cool new methods for subverting the GRE and then just go back to your same old ways when it's time to try practice problems, you won't learn anything. All you'll do is further the same old bad habits.
- Practice under conditions that are as close to the real-life test situation as possible. This means that you should only work when you feel mentally fresh enough to absorb the benefits of what you're doing. If you come home late, don't stay up until the wee hours reading and fighting off yawns. If you can't absorb anything from the process, you're just doing homework for the sake of getting it done.

Other Resources

Keep in mind that there are many other tools available to you so that you can practice all the new techniques you're about to learn.

POWERPREP II Software

The GRE website (**www.ets.org/gre**) has a link to the *POWERPREP II* software. This free program contains a GRE test which you can take on your own computer. It's a great way to get used to the computerized format of the test and try various questions and essay prompts. However, the number of questions it has is very limited, so you should probably save it until you've worked through most of this book.

Books

The most important book (besides this one, of course!) to check out is *The Official Guide to the GRE revised General Test*. This book is published by ETS, and contains questions for every single question type, Math and Verbal, and practice essay prompts. It also contains a CD with a copy of *POWERPREP II* software.

If you want to brush up on your basic math skills, you can also get *Math Smart*, which takes the time to explain, in step-by-step detail, mathematical concepts from the most basic to the most complex.

We at The Princeton Review have other helpful GRE titles to offer you, too, including the *Verbal Workout for the New GRE, 4th Edition* (the sister to this book) and the larger and more comprehensive *Cracking the New GRE, 2012 Edition*. If you're really pressed for time, the short *Crash Course for the New GRE* will give you a quick overview of what you need to know for the test.

On the Web

Books are great learning resources, but they can't replicate the process of working with a computer interface. That's why The Princeton Review has developed several online test-prep resources. At **PrincetonReview.com**, you'll have access to additional lessons and practice GRE tests that will help reinforce what you've learned in the book.

To find out more, surf over to **PrincetonReview.com** or call 1-800-2REVIEW.

Above all: Keep practicing and stay focused. Good luck!

Chapter 2
The End of Mathophobia

WHEN THEY EXPECT YOU TO ZIG, *ZAG*

Before we get into the nuts and bolts of math review, in Chapter 3, we should first discuss what kind of mindset you want to adopt toward the quantitative sections of the GRE. As you know, the GRE is a "standardized" test, in that it tests what ETS believes is a "standard" body of math skills, in what is a mostly standard, multiple-choice format. ETS creates test questions in a standard way, usually by predicting the same, standard errors that most test takers make. Therefore, much of your success on these tests will stem from training yourself to think in a way that ETS hasn't predicted.

In other words, the GRE has set a bunch of traps along the path that most people follow, so you're a lot less likely to fall into them if you take a detour.

STEADY AS SHE GOES

Developing these new ways of thinking—and feeling confident that you've invested so much time and effort into improving your score—can be the greatest asset toward building the most formidable skill possible: *poise.* Poise will help you react calmly when challenges present themselves so that you can search your brain for the right course of action. Poise will help you keep your head about you, when all about you are losing theirs.[1] Poise will make you realize that since people fear what that don't understand, a heightened familiarity with math can cure even the most debilitating case of mathophobia.

THINK NUMERICALLY

First, there's the simple matter of getting used to working with numbers. If you don't come in contact with them very often or if you rely on calculators and/or spreadsheets to do the work for you, it's time to ease your toes back into the numerical hot tub. Even though you'll have a calculator, there will be plenty of times when calculating figures in your head will save you valuable time. So get back into the numerical swing of things.

- Balance your checkbook without using a calculator.
- Figure out a 15 percent tip by calculating 10 percent and adding half again as much (because 10 percent plus 5 percent equals 15 percent).
- Take a few measurements and find out exactly how much storage space that old armoire has.
- Calculate the exact miles per gallon your car got on that last trip to your sister's house.
- Stop by a secondhand book shop and pick up an old math textbook with lots of word problems in it.

1. Say, was that a Kipling reference? In a *math* book? Weird.

Even something as mundane as re-memorizing your times tables from zero to twelve can be useful because it reminds your brain that you were able to speak math fluently not too long ago and that you can do it again. It will also keep you from relying on the calculator for every little math calculation, thus freeing up time for the really important thinking that can be the difference between a good score and a great score.

THINK LIKE A TEST WRITER

Any test writer can write a math question and provide the correct answer. The true "skill" (if you can call it that) of writing questions is in choosing decoy answer choices that an unwitting test taker might choose if he or she made a predictable, careless error like forgetting to carry a 2 or omitting a minus sign.

As you work through all the questions in this book (and in any other resource you use), become acquainted with how GRE questions are written. Take them apart and look under the hood so you can see where you usually make errors and learn how not to make them.

DON'T THINK LIKE ETS THINKS YOU'LL THINK

Quick: Think of a number and write it in this space. _____

If you're like most people, you chose a positive integer, probably between zero and ten. This is the kind of thinking that ETS can anticipate from its test takers, because humans usually think in integers, or "counting numbers." Very few people will respond to the above question by writing -456.49 or $6\frac{14}{61}$ or π or $\sqrt{29{,}534{,}901}$ even though it's just this type of contrary, outside-the-box thinking that often provides good insights to solving math problems.

So when you think about numbers, be sure to consider *all types* of numbers—negatives, fractions, decimals, irrationals, the really huge and the really tiny—because they're as much a part of the number family as boring old integers are.

THINK BACKWARD

Throughout our scholastic lives, we've all been told over and over again to look at a question, show all of our work, and provide our own answer in the blank. There's a reason for this: It's a lot harder to pull an answer out of thin air than to choose the right one from an array of five. A lot of your work on GRE prep will involve taking advantage of this multiple-choice format. So whenever a question makes you feel stuck on questions with multiple answers, remember that the right answer is right there staring back at you. All you have to do is distinguish it from the four (or more!) decoys.

POE SHALL SET YOU FREE

The Process of Elimination (POE) is a beautiful thing, because in many circumstances it will allow you to choose the correct answer without necessarily knowing why it's correct. All you have to know is that four answer choices are definitely wrong in order to know that the answer choice that's left is correct by default.

There is no greater joy for a Princeton Review instructor in the classroom than when a student raises her hand and says, "I got this question correct, but I don't know why." This often means that the student has subverted the system by using POE to choose the correct answer instead of giving up when she didn't know how to approach the problem.

USE THE ANSWER CHOICES AGAINST EACH OTHER

On some questions you'll be able to approximate, or "ballpark," what the correct answer should be and then eliminate the answer choices that are obviously too big or too small. Take this problem, for example:

Question 1 of 20

A dress that normally sells for $100 was marked down by 20%. Then, during a weekend sale, the price was lowered by another 25%. If the dress was sold during that weekend sale, what was its final price, in dollars?

- ○ 45
- ○ 55
- ○ 60
- ○ 120
- ○ 150

Before you attempt anything algebraic—which is an old instinct that you should work hard to eradicate—take a look at the answer choices. If the dress used to sell for $100 and its price was lowered twice, the new price wouldn't be *greater* than $100, would it? So right away you can eliminate (D) and (E).

One important point here: Although on the test the answer choices won't have letters next to them, we'll still refer to them as choices (A) through (E), or (A) through (F), or however many answer choices there are in it.

TURN ALGEBRA INTO ARITHMETIC

As you'll see in Chapter 4, in certain situations you'll be able to plug numbers into the answer choices in order to make them a little easier to analyze. Sometimes you'll be able to avoid algebra altogether and plug the answer choices themselves back into the question. Most questions will be quite vulnerable to these techniques, and it will be in your best interest to exploit this vulnerability.

THINK CONFIDENTLY

Since you're just getting started with this book, you're at a crossroads. At this point, you can become one of the following two types of people:

- those who prepare as much as possible and arrive at the test center knowing they've done all they can, or
- those who arrive at the test center doubting themselves because they know they could have done more to prepare.

Resolve right now to end up in the former category.

But Give Yourself a Break

Students taking the GRE also commonly think too much about their score and become paralyzed with the fear of getting a bad one. This is misdirected energy. Instead, tell yourself that you are going to do the absolute best you can do, and that's all anyone can ask of you—especially yourself.

Once you face down this fear of failure and realize that the world won't end if you don't do well, the fear won't gnaw at you as much and you'll be able to concentrate on the task at hand.

BITE-SIZED PIECES

Whether you like it or not, the quantitative portion of the GRE reads a lot like a verbal section when it comes to word problems. If you're a little intimidated by word problems because of all the information they give you to process, don't consider all of the information at once. Instead, break the question down into bite-size pieces and consider each one separately. Take a look at an example.

Question 4 of 20

Point B is 18 miles east of point A, and point C is 6 miles west of point B. If point D is halfway between points B and C, and point E is halfway between points D and B, how far, in miles, is point E from point D ?

At first glance, this questions looks like an indigestible mouthful of alphabet soup. But you can make sense of it if you're patient and consider it in little sips:

- Point *B* is 18 miles east of point *A:* Draw a line segment with points *A* and *B* on either end.
- Point *C* is 6 miles west of point *B:* Add point *C* to the diagram, 6 miles from *B.*
- Point *D* is halfway between points *B* and *C:* Add point *D,* which is 3 miles from both points *B* and *C.*
- Point *E* is halfway between points *D* and *B:* Add point *E,* which is 1½ miles from both points *B* and *D.*
- And now the question is a piece of cake. **How far, in miles, is point *E* from point *D*?** The answer is 1.5 miles.

See how this kind of step-by-step analysis can help make what looks like an impossible problem relatively simple?

THINK PATIENTLY...

The first time you work on some of these problem types, it may go slowly. That's fine and perfectly normal. When you take the test you'll answer these questions more quickly, but don't worry about that yet. For now, worry about getting the techniques down and doing each question flawlessly. Once you've done that type of question slowly and carefully several times, you will naturally get a little faster at it. Keep striving to answer each question correctly, and speed will come naturally.

BUT DON'T OVER-THINK IT

If you're going to score well on the GRE, naturally you'll have to prepare and practice. But you should also train yourself to manage stress by not over-thinking the process.

Taking the GRE is hard enough; why complicate the process by thinking too much about your score and its consequences? It's too easy to plague yourself with doubts.

Even if you are worried about competitive graduate programs, your GPA, your GRE scores, or intimidating admissions committees, they don't mean anything while you're taking the test. When you're in the middle of a section, your entire universe should be that section only. The past is gone, the future is unknowable. So don't think about either.

THE CALCULATOR

There is a simple on-screen calculator when you take the Math section of the GRE. It doesn't have many functions: plus, minus, multiply, divide, and square root. That's about it. Seems like they're not giving you much, but in reality you don't need that much from a calculator for the GRE. Most of the questions aren't going to require too many calculations, and most of the calculations are straightforward.

> Use the calculator only when necessary.

In fact, before the GRE implemented the on-screen calculator, they tested out how people did with and without the calculator, using unscored Research sections at the end of the test. They found that for most questions, about the same number of people got that question correct with the calculator as did without the calculator. Why? The GRE is not a test of simple calculation. Instead, it tests how well you know how to set up and solve problems.

Feel free to use the calculator for any simple arithmetic you do. Of course, as you do, make sure you write each result on your scratch paper. You will quickly find that you generally don't need to use the calculator more than once or twice per problem, if at all. If you find yourself using the calculator a lot on a single problem, you may either be misreading the problem or doing more work than necessary.

The calculator is simply a tool. It will not solve the questions for you, and it will not replace a good knowledge of basic mathematics. It is better for you to know that $\sqrt{144} = 12$ than it is to pull up the calculator, type in 144, and then hit square root. In fact, the calculator can easily lead you in the wrong direction if you don't know some math rules. For instance, if you type in $5 + 3 \times 2$ into the calculator, you will get an answer of 16, even though the answer is actually 11, due to the order of operations. $5 + 3 \times 2$ is not the sort of math you should use the calculator for.

When to Use the Calculator

Only use the calculator after you have written down the setup for the problem on your scratch paper. Generally, you'll only use it for multiplication and division, and for addition and subtraction of large numbers.

Question 7 of 20

A customer purchased a $300 appliance, which was on sale for 15% off. She paid the sales price plus a sales tax of 10% of the sale price.

Quantity A	Quantity B
The amount, in dollars, the customer paid.	$280

Here's How to Crack It

Since this is a Quant Comp question, write down A B C D vertically on the left side of your paper. We need to take 15% percent of 300, which is the same as 0.15 × 300. Write down 0.15 × 300 on your scratch paper, and then enter it into your calculator. 15% of $300 is $45, so the sale price is $45 less than the original. Write down 300 – 45. Since 300 – 45 is 255, $255 is the sale price. Now we need to calculate the sales tax. 10% of $255 is 0.10 × 255 = 25.50. So the final price, including sales tax, is the sale price plus the sales tax, which is $255 + $25.50 = $280.50. Since Quantity A is larger than Quantity B, the answer is (A).

Notice that before using the calculator, we wrote down the exact formula we were going to enter. Don't keep any information in your head or on the calculator, because in either case there is the risk of losing that information. Write everything down on your scratch paper.

This was a fine question to use the calculator on. The math was fairly straightforward, and we knew how to set up each equation. The calculator simply assisted us with a couple small portions. (If you're not comfortable with how we set up the equations to find each percentage, we'll discuss percentage problems in detail in the next chapter, Nuts and Bolts.)

When NOT to Use the Calculator

There are some questions on the GRE that are created specifically to get you to use the calculator and waste time. If your first impulse is to plug numbers into the calculator, stop. Write down the setup for the problem as usual, and see if there's something else you can do instead. Can you estimate, plug in the answers, or cancel out? Some questions, such as questions involving quadratics, fractions, or exponents, are solvable using a calculator, but take a while unless you look for a different method.

Again, the calculator is not a substitute for knowing the math. If you realize you're writing down really long, ugly decimals, or doing many more steps than usual on the calculator, then you may be missing something with the question.

Question 11 of 20

Amy has a bucket which contains $4\frac{2}{3}$ gallons of water. She pours out $\frac{1}{4}$ of the water. After several hours pass, $\frac{3}{7}$ of the water remaining in the bucket evaporates. How many gallons of water remain?

Give your answer as a fraction.

[ ] gallons

Here's How to Crack It

Your impulse may be to convert all those fractions into decimals. Don't. First off, they're fairly messy as decimals ($\frac{3}{7} \approx 0.42857 \ldots$), and secondly, there are some nice simplifications we can do with the fractions. Start with that $4\frac{2}{3}$ gallons. Convert it into an improper fraction, which means it is $\frac{12}{3} + \frac{2}{3} = \frac{14}{3}$. Now we want to find $\frac{1}{4}$ of $\frac{14}{3}$, and then $\frac{3}{7}$ of that. So we want to find $\frac{14}{3} \times \frac{1}{4} \times \frac{3}{7}$. Notice that we can simplify the 3 in the denominator of the first fraction with the 3 in the numerator of the last fraction, leaving us $\frac{14}{1} \times \frac{1}{4} \times \frac{1}{7}$. Simplify a 7 from the top of the first fraction and the bottom of the last, leaving $\frac{2}{1} \times \frac{1}{4} \times \frac{1}{1}$. Now simplify out a 2 from the first and middle fractions, leaving $\frac{1}{1} \times \frac{1}{2} \times \frac{1}{1} = \frac{1}{2}$. Enter in 1 on the top of the fraction box, and 2 on the bottom.

When You Can't Use the Calculator

The calculator is pretty limited. It can only hold numbers with up to 8 digits, and it can't do exponents, cube roots, or multi-step equations. There will be many problems that simply can't be solved using the calculator. If you know the limits of the on-screen calculator, you'll be less likely to waste time with the calculator on problems you should be solving without it.

Question 15 of 20

If $a = (-5)^{23}$, which of the following is greater than a ?

*Indicate **all** such values.*

- ☐ 2^4
- ☐ $(-3)^{40}$
- ☐ 5^{22}
- ☐ 1
- ☐ $(-6)^{27}$
- ☐ $(-5)^{25}$
- ☐ $(-4)^{16}$
- ☐ $(-5)^{24}$

Here's How to Crack It

Since this is an All That Apply question with 8 answer choices, write down A B C D E F G H. First off, the calculator cannot calculate $(-5)^{23}$. Sorry. It's a gigantic number, much bigger than the calculator could display. Even if it could display it, then you'd still be stuck multiplying $(-5) \times (-5) \times (-5)$ and on and on. There's got to be a shortcut here. As we'll discuss more in detail in Chapter 3: Nuts and Bolts, any negative number raised to an odd exponent is always negative, e.g. $(-2)^3 = (-2)(-2)(-2) = -8$. Any negative number raised to an even exponent is always positive: $(-2)^4 = (-2)(-2)(-2)(-2) = 16$. The exponent on $(-5)^{23}$ is odd, which means that $(-5)^{23}$ is going to be some massively negative number. Since we want to know all the numbers greater than $(-5)^{23}$, all the positive numbers will automatically be larger than our massively negative number. Put check marks next to (A), (B), (C), (D), (G), and (H). All we have left now are (E) and (F). At this point, remember that when dealing with negative numbers, the normal bigger/smaller rules are reversed: –3 is larger than –40. Think of it this way: Which is colder, –10 degrees out or –100 degrees? –100 is much colder, because it's a much smaller number than –10. $(-6)^{27}$ is going to be smaller than $(-5)^{23}$, because 6^{27} is larger than 5^{23}. Cross off (E). The same is true of (F): $(-5)^{25}$ is smaller than $(-5)^{23}$. Cross off (F). The answers are therefore (A), (B), (C), (D), (G), and (H).

Because the calculator is limited to an 8-digit screen, you may occasionally come up against rounding errors for any number which can't fit within the screen. For instance, entering in 0.00001 × 0.00001 on the calculator will give the result as 0, which is definitely wrong. The actual answer is 0.000000001, too small for the calculator to show. Any numbers too large for the calculator to show will result in an Error display.

The calculator is nice to have, but think of it as a backup tool. Make sure you know, without using the calculator, the multiplication table up to 12 × 12, the perfect squares up to 15^2, the cubes up to 5^3, and the decimal and percentage equivalents of most basic fractions. Know that to multiply large numbers ending in zeros, you only have to multiply the non-zero digits then tack on the zeros at the end: 200 × 30 = 6,000. These and other bits of simple math are things you could do with the calculator, but will be much quicker and easier without it.

So let's get to it.

Chapter 3
The Nuts and Bolts

Whenever you decide to learn a new language, what do they start with on the very first day? Vocabulary. Well, math has as much of its own lexicon as any country's mother tongue, so now is as good a time as any to familiarize yourself with the terminology. These vocabulary words are rather simple to learn—or relearn—but they're also very important. Any of the terms you'll read about in this chapter could show up in a GRE math question, so you should know what the test is talking about. (For a complete list, you can consult the glossary in Chapter 13.)

We'll start our review with the backbone of all Arabic numerals: the digit.

DIGITS

You might think there are an infinite number of digits in the world, but in fact there are only ten: 0, 1, 2, 3, 4, 5, 6, 7, 8, and 9. This is the mathematical "alphabet" that serves as the building block from which all numbers are constructed.

Modern math uses digits in a decile system, meaning that every digit in a number represents a multiple of ten. For example, $1{,}423.795 = (1 \times 1{,}000) + (4 \times 100) + (2 \times 10) + (3 \times 1) + (7 \times 0.1) + (9 \times 0.01) + (5 \times 0.001)$.

You can refer to each place as follows:

- 1 occupies the **thousands** place.
- 4 occupies the **hundreds** place.
- 2 occupies the **tens** place.
- 3 occupies the **ones**, or **units**, place.
- 7 occupies the **tenths** place, so it's equivalent to "seven tenths," or $\frac{7}{10}$.
- 9 occupies the **hundredths** place, so it's equivalent to "nine hundredths," or $\frac{9}{100}$.
- 5 occupies the **thousandths** place, so it's equivalent to "five thousandths," or $\frac{5}{1{,}000}$.

When all the digits are situated to the left of the decimal place, you've got yourself an integer.

INTEGERS

When we first learn about addition and subtraction, we start with *integers*, which are the numbers you see on a number line.

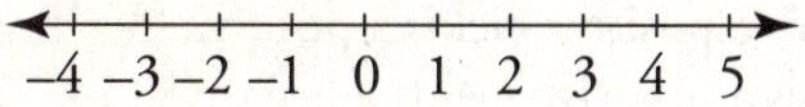

Integers and digits are not the same thing; for example, 39 is an integer that contains two digits, 3 and 9. Also, integers are not the same as **whole numbers,** because whole numbers are only positive. Conversely, integers include negatives *and* zero, and any integer is considered greater than all of the integers to its left on the number line. So just as 5 is greater than 3 (which can be written as $5 > 3$), 0 is greater than –4, and –4 is greater than –10. (For more about greater than, less than, and solving for inequalities, see the next chapter.)

Consecutive Integers and Sequences

Integers can be listed consecutively (such as 3, 4, 5, 6…) or in patterned sequences such as odds (1, 3, 5, 7…), evens (2, 4, 6, 8…), and multiples of 6 (6, 12, 18, 24…). The numbers in these progressions will always get larger, except when explicitly noted otherwise. Note also that, because zero is an integer, a list of consecutive integers that progresses from negative to positive numbers must include it (–2, –1, 0, 1…).

ZERO

Zero is a special little number that deserves your attention. It isn't positive or negative, but it is even. (So a list of consecutive even integers might look like –4, –2, 0, 2, 4….) Zero might also seem insignificant because it's what's called the *additive identity*, which basically means that adding zero to any other number doesn't change anything. (This will be an important consideration when you start plugging numbers into problems in Chapter 4.)

POSITIVES AND NEGATIVES

On either side of zero, you'll find positive and negative numbers. For the GRE, the best thing to know about positives and negatives is what happens when you multiply them together.

- A positive times a positive yields a positive ($3 \times 5 = 15$).
- A positive times a negative yields a negative ($3 \times -5 = -15$).
- A negative times a negative yields a positive ($-3 \times -5 = 15$).

EVEN AND ODD

As you might have guessed from our talk of integers above, even numbers (which include zero) are multiples of 2, and odd numbers are not multiples of 2. If you were to experiment with the properties of these numbers, you would find that

- any number times an even number yields an even number
- the product of two or more odd numbers is always odd
- the sum of two or more even numbers is always even
- the sum of two odd numbers is always even
- the sum of an even number and an odd number is always odd

Obviously, there's no need to memorize stuff like this. If you're ever in a bind, try working with real numbers. After all, if you want to know what you get when you multiply two odd numbers, you can just pick two odd numbers—like 3 and 7, for example—and multiply them. You'll see that the product is 21, which is also odd.

Digits Quick Quiz

Question 1 of 3

If x, y, and z are consecutive even integers and $x < 0$ and $z > 0$, then $xyz =$

Question 2 of 3

If the hundreds digit and ones digit of a three-digit number are interchanged, the new number is 396 less than the old number. Which of the following could be the number?

- ○ 293
- ○ 327
- ○ 548
- ○ 713
- ○ 801

Question 3 of 3

a, b, and c are consecutive digits, and $a > b > c$

Quantity A	Quantity B
abc	$a + b + c$

- ○ Quantity A is greater.
- ○ Quantity B is greater.
- ○ The two quantities are equal.
- ○ The relationship cannot be determined from the information given.

Explanations to Digits Quick Quiz

1. If x, y, and z are consecutive even integers and $x < 0$ and $z > 0$, then x must be –2, y must be 0, and z must be 2. Therefore, their product is 0, and you would enter this number into the box.
2. Take the answer choices and switch the hundreds digit and ones digit. When the result is 396 less than the old number, you have a winner. Answer choices (A), (B), and (C) are out, because their ones digits are greater than their hundreds digits; therefore, the result will be greater (for example, 293 becomes 392). If you rearrange 713, the result is 317, which is 396 less than 713. The answer is (D).
3. Pick three consecutive digits for a, b, and c, such as 2, 3, and 4. Quantity A becomes $2 \times 3 \times 4$, or 24, and Quantity B becomes $2 + 3 + 4$, or 9. Quantity A is greater. But if a, b, and c are –1, 0, and 1, respectively, then both quantities become 0. Therefore, the answer is (D).

PRIME NUMBERS

Prime numbers are special numbers that are only divisible by two distinct factors: themselves and 1. Because neither 0 nor 1 is prime, the smallest prime number is 2. The rest, as you might guess, are odd, because all even numbers are divisible by two. The first ten prime numbers are 2, 3, 5, 7, 11, 13, 17, 19, 23, and 29.

Note that not all odd numbers are prime; 15, for example, is not prime because it is divisible by 3 and 5. Said another way, 3 and 5 are *factors* of 15, because 3 and 5 divide evenly into 15. Let's talk more about factors.

FACTORS AND MULTIPLES

As we said, a prime number has only two distinct factors: itself and 1. But a number that isn't prime—like 120, for example—has several factors. If you're ever asked to list all the factors of a number, the best idea is to pair them up and work through the factors systematically, starting with 1 and itself. So, for 120, the factors are

- 1 and 120
- 2 and 60
- 3 and 40
- 4 and 30
- 5 and 24
- 6 and 20
- 8 and 15
- 10 and 12

Notice how the two numbers start out far apart (1 and 120) and gradually get closer together? When the factors can't get any closer, you know you're finished. The number 120 has 16 factors: 1, 2, 3, 4, 5, 6, 8, 10, 12, 15, 20, 24, 30, 40, 60, and 120. Of these factors, three are prime (2, 3, and 5).

That's also an important point: Every number has a *finite* number of factors.

Prime Factorization

Sometimes the best way to analyze a number is to break it down to its most fundamental parts—its prime factors. To do this, we'll break down a number into factors, and then continue breaking down those factors until we're stuck with a prime number. For instance, to find the prime factors of 120, we could start with the most obvious factors of 120: 12 and 10. (Although 1 and 120 are also factors of 120, because 1 isn't prime, and no two prime numbers can be multiplied to make 1, we'll ignore it when we find prime factors.) Now that we have 12 and 10, we can break down each of those. What two numbers can we multiply to make 12? 3 and 4 work, and since 3 is prime, we can break down 4 to 2 and 2. 10 can be broken into 2 and 5, both of which are prime. Notice how we kept breaking down each factor into smaller and smaller pieces until we were stuck with prime numbers? It doesn't matter which factors we used, because we'll always end up with the same prime factors: $12 = 6 \times 2 = 3 \times 2 \times 2$, or $12 = 3 \times 4 = 3 \times 2 \times 2$. So the prime factor tree for 120 could look something like this:

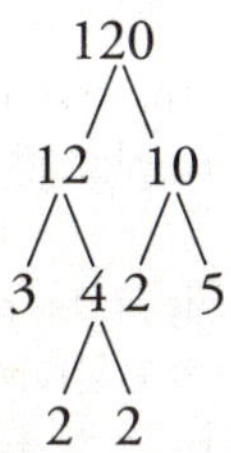

The prime factorization of 120 is $2 \times 2 \times 2 \times 3 \times 5$, or $2^3 \times 3 \times 5$. Note that these prime factors (2, 3, and 5) are the same ones we listed earlier.

MULTIPLES

Because 12 is a factor of 120, it's also true that 120 is a *multiple* of 12. It's impossible to list all the multiples of a number, because multiples trail off into infinity. For example, the multiples of 12 are 12 (12×1), 24 (12×2), 36 (12×3), 48 (12×4), 60 (12×5), 72 (12×6), 84 (12×7), and so forth.

If you ever have trouble distinguishing factors from multiples, remember this:

Factors are Few; Multiples are Many.

Divisibility

If a is a multiple of b, then a is divisible by b. This means that when you divide a by b, you get an integer. For example, 65 is divisible by 13 because $65 \div 13 = 5$.

Divisibility Rules

The most reliable way to test for divisibility is to use the calculator that they give you. If a problem requires a lot of work with divisibility, however, there are several cool rules you can learn that can make the problem a lot easier to deal with. As you'll see later in this chapter, these rules will also make the job of reducing fractions much easier.

1. All numbers are divisible by one. (Remember that if a number is prime, it is only divisible by itself and 1.)
2. A number is divisible by 2 if the last digit is even.
3. A number is divisible by 3 if the sum of the digits is a multiple of 3. For example: 13,248 is divisible by 3 because $1 + 3 + 2 + 4 + 8 = 18$, and 18 is divisible by 3.
4. A number is divisible by 4 if the two digits at the end form a number that is divisible by 4. For example, 13,248 is divisible by 4 because 48 is divisible by 4.
5. A number is divisible by 5 if it ends in 5 or 0.
6. A number is divisible by 6 if it is divisible by both 2 and 3. Because 13,248 is even and divisible by 3, it must therefore be divisible by 6.
7. There is no rule for divisibility by 7.
8. A number is divisible by 8 if the three digits at the end form a number that is divisible by 8. For example, 13,248 is divisible by 8 because 248 is divisible by 8.
9. A number is divisible by 9 if the sum of the digits is a multiple of 9. For example: 13,248 is divisible by 9 because $1 + 3 + 2 + 4 + 8 = 18$, and 18 is divisible by 9.
10. A number is divisible by 10 if it ends in 0.

Remainders

If a number is not divisible by another number, whatever is left over after division is called the *remainder*. You can find the remainder of a division by finding out how many times the smaller number fits into the larger one. For example, when 18 is divided by 5, you know that 5 goes into 18 three times ($3 \times 5 = 15$), and that 3 are left over.

WORKING WITH NUMBERS

A lot of your math calculation on the GRE will require you to know the rules for manipulating numbers using the usual hit parade of mathematical operations: addition, subtraction, multiplication, and division.

PEMDAS (Order of Operations)

When simplifying an expression, you need to perform mathematical operations in a specific order. This order is easily identified by the mnemonic device that most of us come in contact with sooner or later at school—PEMDAS, which stands for Parentheses, Exponents, Multiplication and Division, and Addition and Subtraction. (You might have remembered this as a kid by saying "Please Excuse My Dear Aunt Sally," which is a perfect mnemonic because it's just weird enough not to forget. What the heck did Aunt Sally do, anyway?)

In order to simplify a mathematical term using several operations, perform the following steps:

1. Perform all operations that are in parentheses.
2. Simplify all terms that use exponents.
3. Perform all multiplication and division from left to right. Do not assume that all multiplication comes before all division, as the acronym suggests, because you could get a wrong answer.
   WRONG: $24 \div 4 \times 6 = 24 \div (4 \times 6) = 24 \div 24 = 1$.
   RIGHT: $24 \div 4 \times 6 = (24 \div 4) \times 6 = 6 \times 6 = 36$.
4. Fourth, perform all addition and subtraction, also from left to right.

It's important to remember this order, because if you don't follow it your results will very likely turn out wrong.

Try it out in a GRE example.

Question 3 of 20

$(2 + 1)^3 + 7 \times 2 + 7 - 3 \times 4^2$

Which of the following is the equivalent of the expression above?

- ○ −16
- ○ 0
- ○ 288
- ○ 576
- ○ 1,152

Here's How to Crack It

Simplify $(2 + 1)^3 + 7 \times 2 + 7 - 3 \times 4^2$ like this:

Parentheses:	$(3)^3 + 7 \times 2 + 7 - 3 \times 4^2$
Exponents:	$\mathbf{27} + 7 \times 2 + 7 - 3 \times \mathbf{16}$
Multiply and Divide:	$27 + \mathbf{14} + 7 - \mathbf{48}$
Add and Subtract:	$\mathbf{41} + 7 - 48$
	$\mathbf{48} - 48 = 0$

The answer is (B).

Working with Numbers Quick Quiz

Question 1 of 7

Quantity A	Quantity B
The number of even multiples of 11 between 1 and 100	The number of odd multiples of 22 between 1 and 100

- ○ Quantity A is greater.
- ○ Quantity B is greater.
- ○ The two quantities are equal.
- ○ The relationship cannot be determined from the information given.

Question 2 of 7

42 has the same distinct prime factors as

- ○ 63
- ○ 98
- ○ 210
- ○ 252
- ○ 296

Question 3 of 7

p and r are factors of 100

Quantity A	**Quantity B**
pr	100

- ○ Quantity A is greater.
- ○ Quantity B is greater.
- ○ The two quantities are equal.
- ○ The relationship cannot be determined from the information given.

Question 4 of 7

Quantity A	**Quantity B**
The remainder when 33 is divided by 12	The remainder when 200 is divided by 7

- ○ Quantity A is greater.
- ○ Quantity B is greater.
- ○ The two quantities are equal.
- ○ The relationship cannot be determined from the information given.

Question 5 of 7

$6\ (3 - 1)^3 + 12 \div 2 + 3^2 =$

- ○ 24
- ○ 39
- ○ 63
- ○ 69
- ○ 105

Question 6 of 7

If r, s, t, and u are distinct, consecutive prime integers less than 31, then which of the following could be the average of r, s, t, and u ?

☐ 4

☐ 4.25

☐ 6

☐ 9

☐ 14

☐ 22

☐ 24

Question 7 of 7

The largest prime number that is less than 36 is represented by x. If y represents the smallest even number greater than 19 that is divisible by 3, and x is divided by y, what is the result when 2 is added to that quotient?

☐

Answers to Working with Numbers Quick Quiz

1. The only even multiples of 11 between 1 and 100 are 22, 44, 66, and 88, so Quantity A equals 4. Quantity B is tricky, because if 22 is even, all multiples of 22 are also even. There are no odd multiples of 22, so Quantity B equals 0. The answer is (A).
2. The prime factorization of 42 is $2 \times 3 \times 7$, and those are the distinct prime factors. Answer choice (A) can be eliminated, because 63 is odd. The prime factorization of 252 is $2 \times 2 \times 3 \times 3 \times 7$, so its distinct prime factors are the same. The answer is (D).
3. If p and r are factors of 100, then each must be one of these numbers: 1, 2, 4, 5, 10, 20, 25, 50, or 100. If you plug in $p = 1$ and $r = 2$, for example, then $pr = 2$ and Quantity B is greater. If $p = 50$ and $r = 100$, however, then $pr = 5{,}000$, which is much greater than 100. Therefore, it's impossible to tell if pr is greater than 100, so the answer is (D).
4. 12 goes into 33 two times. The remainder is 9. Since 9 is greater than 7, there is no need to calculate the remainder for column B because it can't possibly be greater than 6. The answer is (A).

5. Follow PEMDAS and calculate the parentheses and exponents first: $6 \times (3 - 1)^3 + 12 \div 2 + 3^2 = 6 \times 8 + 12 \div 2 + 9$. Second, perform all multiplication and division: $6 \times 8 + 12 \div 2 + 9 = 48 + 6 + 9$. Now, it's just a matter of addition: $48 + 6 + 9 = 63$. The answer is (C).
6. This one can be tricky because of the math vocabulary; the question is really asking for the average of four prime numbers in a row. Start by making a list of all the consecutive prime numbers less than 31. Remember that 1 is not prime, and that 2 is the lowest prime number. Your list should be: 2, 3, 5, 7, 11, 13, 17, 19, 23, 29. Starting with 2, 3, 5, 7, use the on-screen calculator to work out the different possible averages of four primes in a row. Choices (A), (C), (E), and (G) would each have to include non-prime numbers and are incorrect. Choices (B), (D), and (F) are the answers.
7. The question is asking for $\frac{x}{y} + 2$, so take it one piece at a time. The largest prime number less than 36 is 31, so $x = 31$. The smallest even number greater than 19 that is divisible by 3 is 24, so $y = 24$. Thus you have $\frac{31}{24} + 2$ and the correct answer is $3.291\overline{6}$ or $\frac{79}{24}$.

PARTS OF THE WHOLE (FRACTIONS)

It's still necessary to be knowledgeable when it comes to fractions, decimals, and percents. Each of these types of numbers has an equivalent in the form of the other two, and fluency among the three of them can save you precious time on test day.

For example: Say you had to figure out 25% of 280. You could take a moment to realize that the fractional equivalent of 25% is $\frac{1}{4}$. At this point, you might see that $\frac{1}{4}$ of 280 is 70, and your work would be done.

If nothing else, memorizing the following table will increase your math IQ and give you a head start on your calculations.

The Conversion Table

Fraction	Decimal	Percent
$\frac{1}{2}$	0.5	50%
$\frac{1}{3}$	$0.33\overline{3}$	$33\frac{1}{3}\%$
$\frac{2}{3}$	0.666	$66\frac{2}{3}\%$
$\frac{1}{4}$	0.25	25%
$\frac{3}{4}$	0.75	75%
$\frac{1}{5}$	0.2	20%
$\frac{2}{5}$	0.4	40%
$\frac{1}{5}$	0.6	60%
$\frac{4}{5}$	0.8	80%
$\frac{1}{6}$	0.166	$16\frac{2}{3}\%$
$\frac{1}{8}$	0.125	12.5%
$\frac{3}{8}$	0.375	37.5%
$\frac{5}{8}$	0.625	62.5%

Fractions

Each fraction is made up of a *numerator* (the number on top) divided by a *denominator* (the number down below). In other words, the numerator is the *part*, and the denominator is the *whole*. By most accounts, the part is less than the whole, and that's the way a fraction is "properly" written.

Improper Fractions

If, in a fraction, the part is greater than the whole, the fraction is considered *improper*. The GRE won't quiz you on the terminology, but it usually writes its multiple-choice answer choices in proper form. A proper fraction takes this form:

$$Integer\ \frac{remainder}{divisor}$$

Converting from Improper to Proper

To convert $\frac{16}{3}$ into proper form, find the remainder when 16 is divided by 3. Because 3 goes into 16 five times with 1 left over, rewrite the fraction by setting aside the 5 as an integer and putting the remainder over the number you divided by (in this case, 3). Therefore, $\frac{16}{3}$ is equivalent to $5\frac{1}{3}$, because 5 is the integer, 1 is the remainder, and 3 is the divisor.

The latter format is also referred to as a *mixed number*, because the number contains both an integer and a fraction.

Converting from Proper to Improper

Sometimes you'll want to convert a mixed number into its improper format. Converting to an improper fraction from a mixed number is a little easier, because all you do is multiply the divisor by the integer, then add the remainder.

$$4\frac{2}{7} = \frac{(4 \times 7) + 2}{7} = \frac{28 + 2}{7} = \frac{30}{7}$$

Improper formats are much easier to work with when you have to add, subtract, multiply, divide, or compare fractions. The important thing to stress here is flexibility; you should be able to work with any fraction the GRE gives you, regardless of what form it's in.

Adding and Subtracting Fractions

There's one basic rule for adding or subtracting fractions: You can't do anything until all of the fractions have the same denominator. If that's already the case, all you have to do is add or subtract the denominators, like this:

$$\frac{3}{13} + \frac{5}{13} = \frac{3+5}{13} = \frac{8}{13}$$

$$\frac{7}{19} - \frac{5}{19} = \frac{7-5}{19} = \frac{2}{19}$$

When the fractions have different denominators, you must convert one or both of them first in order to find their common denominator.

When you were a kid, you may have been trained to follow a bunch of complicated steps in order to find the "lowest common denominator." It might be a convenient thing to learn in order to impress your math teacher, but on the GRE it's way too much work.

The Bowtie

The Bowtie method has been a staple of The Princeton Review's materials since the company began in a living room in New York City in 1981. It's been around so long because it works so simply.

To add $\frac{3}{5}$ and $\frac{4}{7}$, for example, follow these three steps:

Step One: Multiply the denominators together to form the new denominator.

$$\frac{3}{5}+\frac{4}{7}=\frac{\quad}{5\times 7}=\frac{\quad}{35}$$

Step Two: Multiply the first denominator by the second numerator (5 × 4 = 20) and the second denominator by the first numerator (7 × 3 = 21) and place these numbers above the fractions, as shown below.

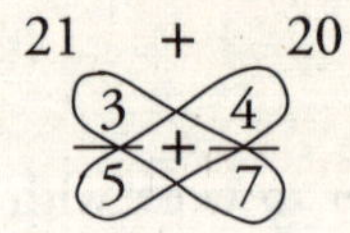

See? A bowtie!

Step Three: Add the products to form the new numerator.

$$\frac{3}{5}+\frac{4}{7}=\frac{21+20}{5\times 7}=\frac{41}{35}$$

Subtraction works the same way.

Note that with subtraction, the order of the numerators is important. The new numerator is 21 – 20, or 1. If you somehow get your numbers reversed and use 20 – 21, your answer will be $-\frac{1}{35}$, which is incorrect. One way to keep your subtraction straight is to always multiply **up** from denominator to numerator when you use the Bowtie.

Multiplying Fractions

Multiplying fractions isn't nearly as complicated as adding or subtracting, because any two fractions can be multiplied by each other *exactly as they are.* All you have to do is multiply all the numerators to find the new numerator, and multiply all the denominators to find the new denominator, like this:

$$\frac{3}{5}\times\frac{4}{7}=\frac{3\times4}{5\times7}=\frac{12}{35}$$

The great thing is that it doesn't matter how many fractions you have; all you have to do is multiply across.

$$\frac{2}{5}\times\frac{3}{8}\times\frac{9}{2}\times\frac{7}{3}=\frac{2\times3\times9\times7}{5\times8\times2\times3}=\frac{378}{240}$$

Dividing Fractions

Dividing fractions is almost exactly like multiplying them, except that you need to perform one extra step:

> When dividing fractions, don't ask why; just flip the second fraction and multiply.

Dividing takes two forms. When you're given a division sign, flip the *second* fraction and multiply them, like this:

$$\frac{2}{5}\div\frac{3}{8}=\frac{2}{5}\times\frac{8}{3}=\frac{2\times8}{5\times3}=\frac{16}{15}$$

Sometimes, you'll be given a compound fraction, in which one fraction sits on top of another, like this:

$$\frac{\frac{4}{7}}{\frac{5}{8}}$$

The fraction bar might look a little intimidating, but remember that a fraction bar is just another way of saying "divide." In this case, flip the *bottom* and multiply.

$$\frac{\frac{4}{7}}{\frac{5}{8}} = \frac{4}{7} \div \frac{5}{8} = \frac{4}{7} \times \frac{8}{5} = \frac{32}{35}$$

Reciprocals of Fractions

A fraction's *reciprocal* is the result when the numerator and denominator of the fraction are "flipped." The reciprocal of $\frac{5}{7}$, for example, is $\frac{7}{5}$.

Furthermore, the product of any number and its reciprocal is 1.

$$\frac{5}{7} \times \frac{7}{5} = \frac{5 \times 7}{7 \times 5} = \frac{35}{35} = 1$$

Knowing this will help you devise a nice shortcut for working with problems like the one on the following page.

Question 8 of 20

Which of the following is equivalent to the reciprocal of $\frac{2\sqrt{3}}{5}$?

- ○ $\frac{10}{\sqrt{3}}$
- ○ $\frac{6}{5\sqrt{3}}$
- ○ $\frac{5\sqrt{3}}{6}$
- ○ $\frac{5}{\sqrt{6}}$
- ○ $\frac{2\sqrt{12}}{10}$

Here's How to Crack It

To solve this problem, multiply each answer choice by the original expression, $\frac{2\sqrt{3}}{5}$. When you get the answer 1, you know you have a match. In this case, the only expression that works is $\frac{5\sqrt{3}}{6}$, so the answer is (C).

Reducing Fractions

Are you scared of reducing, or canceling, fractions because you're not sure what the rules are? If so, there's only one rule to remember:

> You can do anything to a fraction as long as you do exactly the same thing to both the numerator and the denominator.

When you reduce a fraction, you divide both the top and bottom by the same number. If you have the fraction $\frac{3}{15}$, for example, you can divide both the numerator and denominator by a common factor, 3, like this:

$$\frac{3 \div 3}{15 \div 3} = \frac{1}{5}$$

Be Careful

If you have more than one term in the numerator of a fraction but only a single term in the denominator, you can't divide into one of the terms and not the other.

$$\text{Wrong: } \frac{15+8}{4} = \frac{15+\cancel{8}^{\,2}}{\cancel{4}} = \frac{15+2}{1} = 17$$

The only way you can cancel something out is if you can factor out the same number from both terms in the numerator, then divide.

$$\frac{15+3b}{9} = \frac{3\times(5+b)}{3\times 3} = \frac{5+b}{3}$$

Question 13 of 25

Quantity A	Quantity B
$\frac{5}{13}$	$\frac{3}{8}$

○ Quantity A is greater.

○ Quantity B is greater.

○ The two quantities are equal.

○ The relationship cannot be determined from the information given.

Decimals

Decimals are just fractions with a hidden denominator: Each place to the right of the decimal point represents a fraction.

$$0.146 = \frac{1}{10} + \frac{4}{100} + \frac{6}{1000} = \frac{146}{1000}$$

Comparing Decimals

To compare decimals, you have to look at the decimals place by place, from right to left. As soon as the digit in a specific place of one number is greater than its counterpart in the other number, you know which is bigger.

For example, 15.345 and 15.3045 are very close in value because they have the same digits in their tens, units, and tenths places. But the hundredths digit of 15.345 is 4 and the hundredths digit of 15.304 is 0, so 15.345 is greater.

Rounding Decimals

In order to round a decimal, you have to know how many decimal places the final answer is supposed to have (which the GRE will usually specify) and then base your work on the decimal place immediately to the right. If that digit is 5 or higher, round up; if it's 4 or lower, round down.

For example, if you had to round 56.729 to the tenths place, you'd look at the 2 in the hundredths place, see that it was less than 5, and round *down* to 56.7. If you rounded to the hundredths place, however, you'd consider the 9 in the thousandths place and round *up* to 56.73.

Fractions and Decimals Quick Quiz

Question 1 of 8

The fraction $\frac{21}{56}$ is equivalent to what percent?

- ○ 30%
- ○ $33\frac{1}{3}\%$
- ○ 35%
- ○ 37.5%
- ○ 62.5%

Question 2 of 8

What is the sum of $\frac{7}{12}$ and $\frac{2}{5}$?

- ○ $\frac{11}{60}$
- ○ $\frac{11}{17}$
- ○ $\frac{14}{17}$
- ○ $\frac{59}{60}$
- ○ $\frac{59}{17}$

Question 3 of 8

$\left(\frac{1}{27}\right)^{-1}\left(\frac{1}{9}\right)^{-2}\left(\frac{1}{3}\right)^{-3} =$

- ○ 59,049
- ○ 3^{10}
- ○ $\left(\frac{1}{3}\right)^{-10}$

Question 4 of 8

If x is the 32nd digit to the right of the decimal point when $\frac{3}{11}$ is expressed as a decimal, and y is the 19th digit to the right of the decimal point when $\frac{7}{11}$ is expressed as a decimal, what is the value of xy ?

[ ]

Question 5 of 8

If two-thirds of 42 equals four-fifths of x, what is the value of x ?

- ☐ 28
- ☐ 35
- ☐ 42
- ☐ 63
- ☐ 84

Question 6 of 8

Quantity A	**Quantity B**
$\frac{8}{11}$	$\frac{5}{7}$

- ○ Quantity A is greater.
- ○ Quantity B is greater.
- ○ The two quantities are equal.
- ○ The relationship cannot be determined from the information given.

Question 7 of 8

How many digits are there between the decimal point and the first even digit after the decimal point in the decimal equivalent of $\frac{1}{2^8 5^3}$?

Question 8 of 8

A set of digits under a bar indicates that those digits repeat into infinity. What is the value of $(.0000\overline{23})(10^6 - 10^4)$?

- .023
- $.\overline{23}$
- 22.77
- 23
- $23.\overline{23}$

Explanations for Fractions and Decimals Quick Quiz

1. Reduce $\frac{21}{56}$ to $\frac{3}{8}$ and use your conversion table to convert the fraction to 37.5%. The answer is (D).

2. Use the Bowtie method. First, multiply the denominators: $5 \times 12 = 60$. When you multiply diagonally, as in the diagram given in the text, the numerator becomes 35 + 24, or 59. The new fraction is $\frac{59}{60}$, and the answer is (D).

3. Simplify the negative exponents by taking the reciprocal of the corresponding positive exponent, which gives you

$$\left(\frac{1}{\frac{1}{27}}\right)\left(\frac{1}{\left(\frac{1}{9}\right)^2}\right)\left(\frac{1}{\left(\frac{1}{3}\right)^3}\right) = \left(\frac{1}{\frac{1}{27}}\right)\left(\frac{1}{\left(\frac{1}{81}\right)}\right)\left(\frac{1}{\left(\frac{1}{27}\right)}\right)$$

. Now you have

three reciprocals, so flip them over and calculate:

$\left(\frac{27}{1}\right)\left(\frac{81}{1}\right)\left(\frac{27}{1}\right)$ = 59, 049. You can also express each number as a power of 3, which gives you (33)(34)(33) = 310 , which makes choice (B) correct. 310 can also be expressed as $\left(\frac{1}{3}\right)^{-10}$. Thus, the correct answer is choices (A), (B), and (C).

4. Divide to convert the fractions into decimals. First, $\frac{3}{11}$ = .2727 repeating. This is really a pattern question: The odd numbered terms are 2 and the even numbered terms are 7. The 32nd digit to the right of the decimal is an even term, so $x = 7$. Next, $\frac{7}{11}$ = .6363 repeating. This time, the odd numbered terms are 6 and the even numbered terms are 3. The 19th digit on the right side of the decimal place is an odd term, so $y = 6$. Lastly, $xy = 7 \times 6 = 42$.

5. Multiply both sides by the reciprocal, $\frac{5}{4}$ to eliminate the fraction and isolate the x. The fractions on the right cancel out and on the left you have $(28)(\frac{5}{4}) = 35$, so $35 = x$. The answer is (B).

6. To compare two fractions, just cross multiply and compare. $7 \times 8 = 56$ and $11 \times 5 = 55$, so both fractions are very close to even, but Quantity A is greater.

7. The answer is 0. First, divide to convert $\frac{1}{2^8 5^3}$ into a decimal. The on-screen calculator doesn't do exponents, so you may want to factor the denominator: $\frac{1}{2^4 2^4 5^3} = \frac{1}{16 \times 16 \times 125} = \frac{1}{32{,}000} = 0.00003125$. Since 0 is even, there are no digits between the decimal point and the first even digit after the decimal point, and the correct answer is 0. In fact, if you noticed that any decimal starting with a 0 would have the same answer, you only needed to make sure the denominator was larger than 10.

8. Before you break out the calculator, think of distributing the $(0.0000\overline{23})$. This gives you $(0.0000\overline{23})(10^6) - (0.0000\overline{23})(10^4)$. Now the move the decimal point 6 places to the right for the first term and 4 places to the right for the second term, which gives you $23.\overline{23} - .\overline{23} = 23$. The correct answer is choice (D).

Percents

As you may have noted from the conversion chart, decimals and percents look an awful lot alike. In fact, all you have to do to convert a decimal to a percent is to move the decimal point two places to the right: 0.25 becomes 25%, 0.01 becomes 1%, and so forth. This is because they're both based on multiples of 10. Percents also represent division with hidden denominators, but for percents, the denominator is always 100.

Calculating Percents

There are four different ways to calculate percents. Some people find that one way makes more sense than others. There is no best way to find percents. Try out all four, and figure out which one seems most natural for you. You will probably find that some methods work best for some problems, and other methods work best for others.

Translating

Translating is one of the more straightforward and versatile methods of calculating percents. Each word in a percent problem is directly translated into a mathematical term, according to the following chart:

Term	Math equivalent
what	x (variable)
is	=
of	· (multiply)
percent	÷ 100

This table can be a great help. For example, take a look at this Quant Comp problem.

Question 7 of 20

Quantity A	Quantity B
24% of 15% of 400	52% of 5% of 600

- ○ Quantity A is greater.
- ○ Quantity B is greater.
- ○ The two quantities are equal.
- ○ The relationship cannot be determined from the information given.

Here's How to Crack It

Since it's not immediately obvious which quantity is larger, we're going to have to do some actual calculation. Let's start with Quantity A. 24% of 15% of 400 is can be translated, piece by piece, into math. Remember that % means to divide by 100 and *of* means to multiply. As always, write down everything on your scratch paper. After translation, we get $\frac{24}{100}\times\frac{15}{100}\times\frac{400}{1}$. Cancel out one of the 100s in the denominator with the 400 in the numerator of the last fraction to get $\frac{24}{100}\times\frac{15}{1}\times\frac{4}{1}$. Now feel free to use the calculator. Multiply the top of the fractions first, then the bottom, giving you $\frac{1440}{100}=14.4$. Now let's do the same with Quantity B. 52% of 5% of 600 becomes $\frac{52}{100}\times\frac{5}{100}\times\frac{600}{1}$. Cancel out the first 100 with the 600 and simplify $\frac{5}{100}$ to $\frac{1}{20}$, and you'll have $\frac{52}{1}\times\frac{1}{20}\times\frac{6}{1}=\frac{312}{20}=15.6$. Quantity B is therefore bigger, and the answer is (B).

Word Problems and Percents

Word problems are also far less onerous when you apply the math translation table.

Question 16 of 20

Thirty percent of the graduate students at Hardcastle State University come from outside the state, and 75% of out-of-state tuitions receive some sort of financial aid. If there are 3,120 graduate students at Hardcastle State, how many out-of-students do NOT receive financial aid?

- ○ 234
- ○ 468
- ○ 702
- ○ 936
- ○ 1,638

Here's How to Crack It

Note the range in answer choices. This is another problem that is ripe for ballparking. Assume there are 3,000 graduate students. 10% of 3,000 is 300, so 30% is 900. 25% do not receive financial aid, which is what you're looking for, so ¼ of 900 is 225. Since there are few more than 3,000 graduate students our answer must be a few more than 225. Only answer (A) is even close. Note that answer choice (C) represents the 75% who do receive financial aid. Make sure you read slowly and carefully and take all problems in bite sized pieces.

Conversion

The second way to deal with percentage questions is to use the chart on page 42. This will allow you to quickly change each percentage into a fraction or a decimal. This method works well in tandem with Translating, as shown in the explanation for Question 7, above. Since we knew that $\frac{5}{100} = \frac{1}{20}$, we were able to save ourselves some multiplication with the calculator.

Proportions

You can also set up a percent questions as a proportion by matching up the part and whole. (Some people learn this method as "is over of.") Let's look at a couple quick examples:

1) 60 is what percent of 200?
Because 60 is some part of 200, we can set up the following proportion:

$$\frac{60}{200} = \frac{x}{100}$$

Notice that each fraction is simply the part divided by the whole. On the left side, 60 is part of 200. We want to know what that part is in terms of a percent, so on the right side we set up *x* (the percent we want to find), divided by 100 (the total percent).

2) What is 30% of 200?
Now we don't know how much of 200 we're dealing with, but we know the percentage. So we'll put our unknown, *x*, over the whole, 200, and our percentage on the right:

$$\frac{x}{200} = \frac{30}{100}$$

3) 60 is 30% of what number?
We know our percentage, and we know our part, but we don't know the whole, so that's our unknown:

$$\frac{60}{x} = \frac{30}{100}$$

Notice that the setup for each problem (they're actually just variations of the same problem: 30% of 200 is 60) is essentially the same. We had one unknown, either the part, the whole, or the percentage, and we wrote down everything we knew.

Every proportion will therefore look like:

$$\frac{part}{whole} = \frac{percentage}{100}$$

Once you've set up the proportion, cross multiply to solve.

Question 17 of 20

Ben purchased a computer and paid 45% of the price immediately. If he paid $810 immediately, what is the total price of the computer?

Here's How to Crack It

Let's find the parts of the proportion that we know. We know he paid 45%, so we know the percentage, and we know the part he paid: $810. We're missing the whole price. On your scratch paper, set up the proportion $\frac{810}{x} = \frac{45}{100}$. Cross multiply, and you'll get $45x = 81{,}000$. Divide both sides by 45 (feel free to use the calculator here), and you'll get $x = \$1{,}800$.

Tip Calculation

The last method for calculating percentages is a variation on a method many people use to calculate the tip for a meal.

To find 10% of any number, simply move the decimal one place to the left.

10% of 100 = 10.0
10% of 30 = 3.0
10% of 75 = 7.5
10% of 128 = 12.8
10% of 87.9 = 8.79

To find 1% of any number, move the decimal place two places to the left.

1% of 100 = 1.00
1% of 70 = 0.70
1% of 5 = 0.05
1% of 2,145 = 21.45

You can then find the value of any percentage by breaking the percentage into 1%, 10%, and 100% pieces. Remember that 5% is half of 10%, and 50% is half of 100%.

5% of 60 = half of 10% = half of 6 = 3
20% of 35 = 10% + 10% = 3.5 + 3.5 = 7
52% of 210 = 50% + 1% + 1% = 105 + 2.1 + 2.1 = 109.2
40% of 70 = 10% + 10% + 10% + 10% = 7 + 7 + 7 + 7 = 28

Generally, we'll use this most often with Ballparking, especially in Charts and Graphs questions.

Question 18 of 20

Last month Dave spent at least 20% but no more than 25% of his monthly income on groceries. If his monthly income was $2,080.67, which of the following could be the amount Dave spent on groceries?

*Indicate **all** such values.*

☐ $391.92

☐ $432.88

☐ $456.02

☐ $497.13

☐ $530.17

☐ $545.60

☐ $592.43

Here's How to Crack It

Ugh. Those are some ugly numbers. Let's Ballpark a little bit, and use some quick tip calculations to simplify. First, write down A B C D E F G vertically on your scratch paper. We're looking for a number bigger than 20% of $2,080.67. Ignore the 67 cents for now. If any answers are only a couple pennies away from our answer, then we can go back and use more exact numbers, but that's fairly unlikely. 10% of $2,080 is $208, which means that 20% is $208 + $208 = $416. Dave must have spent at least $416 on groceries, so cross off answer (A). We know he spent no more than 25% of his income. 5% of $2,080 is half of 10%, which means that 5% of $2,080 is $104, and 25% of $2,080 is 10% + 10% + 5% = $208 + $208 + $104 = $520. He couldn't have spent more than $520 on groceries, so cross off answers (E), (F), and (G). The answers are everything between $416 and $520: (B), (C), and (D).

Percent Change

As we explored in the introduction, percent change is based on two quantities: the change and the original amount.

Trigger: A question asks for "percent change," "percent increase," or "percent decrease."

Response: Write the percentage change formula.

$$\% \, change = \frac{change}{original} \times 100$$

Question 19 of 20

Quantity A	Quantity B
The percent change from 10 to 11	The percent change from 11 to 10

- ○ Quantity A is greater.
- ○ Quantity B is greater.
- ○ The two quantities are equal.
- ○ The relationship cannot be determined from the information given.

Here's How to Crack It

At first glance, you might assume that the answer is (C), because the numbers in each quantity look so similar. However, even though both quantities changed by a value of 1, the original amounts are different. When you follow the formula, you find that Quantity A equals $\frac{1}{10} \times 100$, or 10%, while Quantity B equals $\frac{1}{11} \times 100$, or 9%. The answer is (A).

Percentage change also factors into a lot of "real-world math," so we'll talk about it more in Chapter 5. From here, we head to a powerful little device that helps us convey very big and very small numbers with very little effort: exponents. Before we get into exponents, though, let's practice those percents.

Percents Quick Quiz

Question 1 of 6

What is 26% of 3,750 ?

- ○ 753
- ○ 975
- ○ 1,005
- ○ 2,775
- ○ 9,750

Question 2 of 6

Quantity A	Quantity B
200% of 20% of 300	120% of 25% of 400

- ○ Quantity A is greater.
- ○ Quantity B is greater.
- ○ The two quantities are equal.
- ○ The relationship cannot be determined from the information given.

Question 3 of 6

A think tank projects that the people over the age of 65 will comprise 25% of the U.S. population by 2010. If the current population of 300 million is expected to grow by 8% by 2010, how many people over 65, in millions, will there be?

[ ]

Question 4 of 6

Marat sold his condo at a price that was 18% more than the price he paid for it. If he bought his condo at price that was 42% less than the buyer's $175,000 asking price, which of the following must be true?

- ☐ The person who sold Marat the condo lost money.
- ☐ Marat bought the condo for $101,500.
- ☐ 36 percent of the price at which Marat sold the condo is $43,117.20.

Question 5 of 6

If 25% of p equals 65% of 80 and if q is 50% of p, which of the following must be true?

☐ 65 is 62.5% of q

☐ q is 130% of 80

☐ p is 200% of q

☐ $\frac{q}{p}$ = 50% of 1

Question 6 of 6

If John withdraws x% of his $13,900 balance from his checking account, his new balance will be less than $10,000.

Quantity A	**Quantity B**
x	25

○ Quantity A is greater.

○ Quantity B is greater.

○ The two quantities are equal.

○ The relationship cannot be determined from the information given.

Explanations for Percents Quick Quiz

1. Convert 26% is close to 25%, so what is $\frac{1}{4}$ of 3,750? Only answer choice (B) is close.

2. In Quantity A, 20% or $\frac{1}{5}$ of 300 is 60. 100% of 60 is 60, so 200% must be 120. In Quantity B, 25% or $\frac{1}{4}$ of 400 is 100. 20% of 100 is 20, so 120% is 120. The two quantities are equal, and the answer is (C).

3. 1% of 300 million is 3 million, so 8% is 24 million. If the population grew by 8% that means that it added another 24 million people, so the new total is 324 million. 25% or $\frac{1}{4}$ of 324 million is 81. So there will be approximately 81 million people over 65 in the U.S. in 2010.

4. Take bite-sized pieces. Marat bought his condo for 58 percent of \$175,000 = \$101,500. (Alternatively, you could take 42 percent of \$175,000 and subtract that from \$175,000 to get \$101,500.) Choice (B) is correct. Marat later sold his condo for 18 percent more than he paid for it. 1.18 × \$101,500 = 119,770. 36 percent of that selling price is 0.36 × 119,700 = \$43,117.20. Choice (C) is correct. Because you only know the asking price of the person who sold Marat the condo, and not how much he or she paid for it, choice (A) may or may not be true. Eliminate it. The correct answers are choices (B) and (C).

5. Start by translating: $0.25p = 0.65(80)$, so $0.25p = 52$ and $p = 208$. That means $q = 0.50(208) = 104$. Now replace p and q in the answer choices with the appropriate values. For choice (A), you have $65 = \frac{62.5}{100} \times 104$, which is true. Choice (A) is correct. For choice (B), $104 = \frac{130}{100} \times 80$. This equation is also true, so keep choice (B). Choice (C) is also correct because $208 = \frac{200}{100} \times 104$. Finally, choice (D) is correct because $\frac{104}{208} = \frac{1}{2}$, which is 50 percent of 1. The correct answers are choices (A), (B), (C), and (D).

6. Whenever you see a variable in one column and an actual number in the other column, try plugging the number in for the variable. In this case, if John withdrew 25% of his savings his balance would still be \$10,425. Therefore he must withdraw more than 25% in order for his balance to dip below \$10,000. The answer is (A).

EXPONENTS

The small number to the upper right corner of an integer or other math term is called an exponent, and it tells you how many times that variable should be multiplied by itself. For example, $5^4 = 5 \times 5 \times 5 \times 5$. The exponent is 4, and the base is 5.

You can add or subtract two exponential terms as long as they're like terms.

$5x^5 + x^5 = 6x^5$

$6b^3 - 4b^3 = 2b^3$

$15ab^2c^3 - 9ab^2c^3 + 2ab^2c^3 = 8ab^2c^3$

Multiplying and Dividing Exponential Terms

Let's say you're multiplying $a^2 \times a^3$. If we expand out those fractions, then we get $(a \times a)(a \times a \times a) = a^5 = a^{(2+3)}$. So when multiplying terms that have the same base, you add the exponents.

Note that the terms have to have the same base. If we're presented with something like $a^2 \times b^3$, then we can't simplify it any further than that.

Now let's deal with division. Let's start by expanding out an exponent problem using division, to see what we can eliminate.

$$\frac{a^5}{a^3} = \frac{a \times a \times a \times a \times a}{a \times a \times a} = \frac{a \times a}{1} = a^2$$

Notice that the three a terms on the bottom canceled out with three a terms on top? We ended up with a^2, which is the same as $a^{(5-3)}$. So when dividing terms that have the same base, subtract the exponents.

Parentheses with exponents work exactly as they do with normal multiplication. $(ab)^3 = (ab)\ (ab)\ (ab) = a^3b^3$. When a term inside parentheses is raised to a power, distribute the exponent to each individual term within the parentheses.

What if a term inside the parentheses already has an exponent? For example, what if we have $(a^2)^3$? Well, that's $(a^2)\ (a^2)\ (a^2) = (a \times a)\ (a \times a)\ (a \times a) = a^6 = a^{2\times3}$. So when you raise a term with an exponent to another power, multiply the two exponents.

To review, the big rules with exponents are:

- When **Multiplying** terms with the same base, **Add** the exponents.
- When **Dividing** terms with the same base, **Subtract** the exponents.
- When raising a term with an exponent to another **Power, Multiply** the exponents.

Most exponent questions will use one or more of these rules. You can memorize them as **MADSPM**: Multiply, Add, Divide, Subtract, Power, Multiply.

When in Doubt, Expand It Out

If you ever have trouble remembering any of these rules, you can always fall back on a very valuable guideline: "When in doubt, expand it out." Here's an example of how this works:

Question 8 of 20

Which of the following is equivalent to $(2x)^3 \cdot 5x^2 \cdot (x^4)^5$?

- ○ $10x^{15}$
- ○ $20x^{15}$
- ○ $20x^{25}$
- ○ $40x^{15}$
- ○ $40x^{25}$

Here's How to Crack It

When you expand everything out, the factors look like this:

$$(2x)^3 \cdot 5x^2 \cdot (x^4)^5 = [(2x) \cdot (2x) \cdot (2x)] \cdot [5 \cdot x \cdot x] \cdot [(x^4) \cdot (x^4) \cdot (x^4) \cdot (x^4) \cdot (x^4)]$$

$$= 2 \cdot 2 \cdot 2 \cdot 5 \cdot x^1 \cdot x^1 \cdot x^1 \cdot x^1 \cdot x^1 \cdot x^4 \cdot x^4 \cdot x^4 \cdot x^4 \cdot x^4$$

$$= 40 \cdot x^{(1+1+1+1+1+4+4+4+4+4)} = 40x^{25}$$

The answer is (E).

Rules, Quirks, Anomalies, and Other Weirdness

There are several other general peculiarities about exponential terms that you should at least appreciate.

- Any number raised to the first power equals itself: $x^1 = x$.
- Any number raised to the zero power equals one: $x^0 = 1$.
- Raising a negative number to an even power results in a positive number: $(-2)^4 = 16$.
- Raising a negative number to an odd power results in a negative number: $(-2)^5 = -32$.
- Raising a fraction between 0 and 1 to a power greater than one results in a smaller number: $\left(\frac{1}{2}\right)^2 = \frac{1}{2} \times \frac{1}{2} = \frac{1}{4}$.
- Finding a root of a fraction between 0 and 1 results in a larger number: $\sqrt{\frac{1}{4}} = \frac{\sqrt{1}}{\sqrt{4}} = \frac{1}{2}$.

Comparing Exponential Terms

Comparing two exponential terms is easy if all of the numbers are positive and the two terms have the same base. Just by looking, you can determine that 3^5 is greater than 3^4, because 5 is greater than 4. So what happens when the bases are different?

Question 9 of 20

Quantity A	Quantity B
64^5	16^8

- ○ Quantity A is greater.
- ○ Quantity B is greater.
- ○ The two quantities are equal.
- ○ The relationship cannot be determined from the information given.

Trigger: Exponent problem with large numbers.

Response: Factor the base to compare exponents.

Here's How to Crack It

Holy smokes. Those two numbers sure look huge, don't they? And they are, but lucky for you their exact value isn't important. All you have to do is compare them. So resist the temptation to plug 64 × 64 × 64 × 64 × 64 into your calculator.

That's exactly what ETS wants you to do, because it's clumsy and time-consuming. The way to improve your math score is to recognize patterns that *save* time.

So which number is larger? Is it Quantity A, which has the larger base? Or is it Quantity B, which has the larger exponent? Let's find out.

Both 16 and 64 are multiples of 4. In fact, $16 = 4 \times 4$, or 4^2, and $64 = 4 \times 4 \times 4$, or 4^3. Therefore, you can rewrite each of the quantities above using 4 as the common base, and if you apply your newfound knowledge of exponential rules, the comparison will become much more apparent:

$$64^5 = (4^3)^5 = 4^{3 \times 5} = 4^{15}$$

$$16^8 = (4^2)^8 = 4^{2 \times 8} = 4^{16}$$

Do we care how much either of those two quantities is? Absolutely not. Because all the numbers are positive, it must be true that 4 raised to the greater power is the larger number. Therefore, the answer is (B).

Negative Exponents

Any number raised to a negative exponent can be rewritten in reciprocal format with a positive exponent: $x^{-3} = \frac{1}{x^3}$, so $8^{-3} = \frac{1}{8^3} = \frac{1}{512}$. You can analyze a Quant Comp question that involves negative exponents in much the same way.

Question 4 of 20

Quantity A	Quantity B
27^{-4}	9^{-8}

- ○ Quantity A is greater.
- ○ Quantity B is greater.
- ○ The two quantities are equal.
- ○ The relationship cannot be determined from the information given.

Here's How to Crack It

Because 27 and 9 are both multiples of 3, you can rewrite each quantity using 3 as the common base.

$$27^{-4} = (3^3)^{-4} = 3^{3 \times -4} = 3^{-12}$$

$$9^{-8} = (3^2)^{-8} = 3^{2 \times -8} = 3^{-16}$$

Again, we don't care about the actual values of these numbers, which are very, very small. All we need to do is compare the exponents: Because –12 is greater than –16, the answer is (A).

Scientific Notation

In real life, the chief purpose of exponents is to express humongous or teeny-tiny numbers conveniently. The sciences are chock full of these sorts of numbers—such as Avogadro's number (6.022×10^{23}), which helps chemistry students determine the molecular weight of each element in the periodic table. On the GRE, you probably won't see Avogadro's number, but you may well see a number expressed in scientific notation. Scientific notation is basically a number multiplied by 10 raised to a positive or negative power.

Scientific notation merely serves to make unwieldy numbers a little more manageable. If you see one on the GRE, just remember these rules:

- If the exponent is positive, then move the decimal point that many spaces to the right ($6.022 \times 10^{23} = 602{,}200{,}000{,}000{,}000{,}000{,}000{,}000$); and
- If the exponent is negative, then move the decimal point that many spaces to the left ($4.5 \times 10^{-15} = 0.0000000000000045$).

Fractional Exponents

Any number raised to a fractional exponent can be rewritten in root format: $x^{\frac{1}{3}} = \sqrt[3]{x}$ (or "the cube root of x"), so $8^{\frac{1}{3}} = \sqrt[3]{8} = 2$.

Question 11 of 20

Quantity A	Quantity B
$16^{\frac{1}{2}}$	$\left(\frac{1}{2}\right)^{-2}$

- ○ Quantity A is greater.
- ○ Quantity B is greater.
- ○ The two quantities are equal.
- ○ The relationship cannot be determined from the information given.

Here's How to Crack It

Quantity A can be rewritten as $\sqrt{16}$, which equals 4. Quantity B is a little trickier, but you can figure it out if you follow a few of the rules we've discussed in this chapter.

$$\left(\frac{1}{2}\right)^{-2} = \frac{1}{\left(\frac{1}{2}\right)^2} = \frac{1}{\frac{1}{4}} = 1 \times \frac{4}{1} = 4$$

Both quantities are equal to 4, so the answer is (C).

What's that? Never heard of roots? Well, roots are covered in the next section, right after the quick quiz. How's that for an impressive segue?

Exponents Quick Quiz

Question 1 of 4

What is the product of $2a^3$ and $5a^7$?

- ○ $7a^{10}$
- ○ $7a^{21}$
- ○ $10a^{10}$
- ○ $10a^{21}$
- ○ $14a^{15}$

Question 2 of 4

Quantity A	Quantity B
$200x^{295}$	$10x^{294}$

- ○ Quantity A is greater.
- ○ Quantity B is greater.
- ○ The two quantities are equal.
- ○ The relationship cannot be determined from the information given.

Question 3 of 4

Quantity A	Quantity B
$3^{-2} + 4^{-2}$	$\left(\frac{5}{12}\right)^2$

- ○ Quantity A is greater.
- ○ Quantity B is greater.
- ○ The two quantities are equal.
- ○ The relationship cannot be determined from the information given.

Question 4 of 4

Which of the following is equivalent to $144^{\frac{1}{2}} \div 2^2$?

- ○ $27^{\frac{1}{3}}$
- ○ $2^3 \times 3^2$
- ○ $64^{\frac{1}{6}}$
- ○ $8^{\frac{1}{3}} + 4^{\frac{1}{2}}$
- ○ $4^{\frac{1}{2}} \times 16^{\frac{1}{2}}$

Explanations for Exponents Quick Quiz

1. Multiply the coefficients first: $2 \times 5 = 10$. When you multiply the exponential terms, add 3 and 7 to get 10. The combined term is $10a^{10}$, and the answer is (C).
2. This is a tricky one, because it looks like Quantity A will always be larger due to the greater exponent. If x is negative, however, it's a different story. Because –1 to any odd power is negative and –1 to any even power is positive, Quantity B becomes larger. The answer is (D).
3. Quantity A translates to $\frac{1}{3^2}+\frac{1}{4^2}$, or $\frac{1}{9}+\frac{1}{16}$. You can use the Bowtie to add these and get $\frac{25}{144}$, which happens to equal $\left(\frac{5}{12}\right)^2$. The answer is (C).
4. An exponent of $\frac{1}{2}$ is equivalent to a square root, so you can rewrite the term as $\sqrt{144} \div 4$, which equals $12 \div 4$, or 3. An exponent of $\frac{1}{3}$ is equivalent to a cube root, and because 3 is the cube root of 27, the answer is (A).

ROOTS

A square root denoted by the funny little check mark with an adjoining roof: $\sqrt{\ }$. The number inside the house is called a *radicand.*

Square roots can cause a lot of confusion and despair because it can be hard to remember when you can combine them and when you can't. As a result, you see people adding square roots like this:

$$\text{WRONG: } \sqrt{4}+\sqrt{9}=\sqrt{13}$$

This is absolutely wrong, and it's easy to prove it: $\sqrt{4}=2$ and $\sqrt{9}=3$, so the left side of the equation can be rewritten as 2 + 3, or 5. 5 is equal to $\sqrt{25}$, not $\sqrt{13}$.

If you ever find yourself in a jam when it comes to remembering square-root rules, try fiddling around with some numbers as we did in that last example. You'll start to see which manipulations of square roots are legal and which aren't, and that will help you understand the rules more organically.

Perfect Squares

If the square root of a number is an integer, then that number is known as a *perfect square*. Perfect squares will come up a lot in the rest of the chapter, and it pays to be able to recognize them on sight. The first ten perfect squares are 1, 4, 9, 16, 25, 36, 49, 64, 81, and 100 (and it couldn't hurt if you added the next five—121, 144, 169, 196, and 225—to your list).

Knowing your perfect squares is a great tool to use when estimating. Because 70 is between 64 and 81, for example, it must be true that $\sqrt{70}$ is between 8 and 9, because $\sqrt{64} = 8$ and $\sqrt{81} = 9$.

The on-screen calculator can find any square root you're unsure of, but get in the habit of being able to estimate square roots for numbers up to 200. It'll save you time, since you won't have to keep bringing up the calculator, and it'll help you realize when you've made a mistake entering anything in your calculator. Feel free to go back to the calculator if you're not sure, but learn those perfect squares.

Question 19 of 20

How many even integers are there between $\sqrt{50}$ and $\sqrt{150}$?

- ○ 3
- ○ 5
- ○ 10
- ○ 50
- ○ 100

Here's How to Crack It

7^2 is 49. 8^2 is 64. The $\sqrt{50}$, therefore, will be a number just larger than 7 but much less than 8. 12^2 is 144, and 13 squared is 169. The $\sqrt{150}$, therefore, is a number just larger than 12 but a lot less than 13. Count the even integers between 7 and 12, including the 12: 8, 10, 12. The answer is (A).

Multiplying Roots

Multiplying two square roots is a rather straightforward process; you just multiply the numbers and put the result under a new square root sign.

$$\sqrt{15} \times \sqrt{3} = \sqrt{15 \times 3} = \sqrt{45}$$

If there are numbers both outside and inside the square-root sign, you multiply them separately and then put them all together.

$$6\sqrt{15} \times 2\sqrt{3} = (6 \times 2) \times \sqrt{15 \times 3} = 12\sqrt{45}$$

Dividing Roots

Dividing roots involves much the same process, but in reverse. When you divide roots, it's often helpful to set the division up as a fraction, like this:

$$\sqrt{15} \div \sqrt{3} = \frac{\sqrt{15}}{\sqrt{3}} = \sqrt{\frac{15}{3}} = \sqrt{5}$$

Again, if you have numbers both outside and inside the square root sign, divide them separately, like this:

$$6\sqrt{15} \div 2\sqrt{3} = \frac{6\sqrt{15}}{2\sqrt{3}} = \frac{6}{2} \times \frac{\sqrt{15}}{\sqrt{3}} = 3 \times \sqrt{\frac{15}{3}} = 3\sqrt{5}$$

Simplifying Roots

If the radicand has a factor that is a perfect square, the term can be simplified. The GRE doesn't ask you to do this very often, but it pays to have the ability, just in case your answer is not in its simplest form but the answer choices are.

When the two expressions were multiplied in the above example, the answer included the term $\sqrt{45}$. Because 45 = 9 × 5, and 9 is a perfect square, you can simplify the expression using the rules of multiplication that you just learned. In fact, you just follow the directions in reverse order.

$$\sqrt{45} = \sqrt{9 \times 5} = \sqrt{9} \times \sqrt{5} = 3\sqrt{5}$$

If there's already a number sitting outside the square-root sign, you will need to combine it with the simplified term like this:

$$12\sqrt{45} = 12 \times \sqrt{9 \times 5} = 12 \times \sqrt{9} \times \sqrt{5} = 12 \times 3 \times \sqrt{5} = 36\sqrt{5}$$

Adding and Subtracting Roots

You can add and subtract square roots just like variables as long as they have the same radicands.

$$5\sqrt{13} + 3\sqrt{13} = 8\sqrt{13}$$

$$9\sqrt{6} - 2\sqrt{6} = 7\sqrt{6}$$

If the radicands are different, however, you can't do a thing with them; $\sqrt{6} + \sqrt{5}$, for example, cannot be combined or simplified. The only way you can hope to combine two square roots that have different radicands is if you can combine them. The key is to determine if the radicands have factors that are perfect squares.

For example, look at the expression $2\sqrt{12}+\sqrt{75}$:

$$2\sqrt{12}=2\times\sqrt{4\times 3}=2\times\sqrt{4}\times\sqrt{3}=2\times 2\times\sqrt{3}=4\sqrt{3}$$

$$\sqrt{75}=\sqrt{25\times 3}=\sqrt{25}\times\sqrt{3}=5\times\sqrt{3}=5\sqrt{3}$$

Therefore, $2\sqrt{12}+\sqrt{75}$ can be rewritten as $4\sqrt{3}+5\sqrt{3}$, which equals $9\sqrt{3}$.

Rationalizing Roots

There's a rule in lots of moldy old math textbooks that says you have to rationalize a square root in the denominator of a fraction. The GRE doesn't enforce this rule, but you should be able to recognize equivalent values of fractions with square roots in them. (This is especially important for geometry questions, as we'll discuss in Chapter 7.) In order to rationalize a fraction with a square root in its denominator, you need to multiply both the numerator and the denominator by the square root.

To rationalize $\frac{1}{\sqrt{2}}$, for example, watch what happens when you multiply the top and bottom by $\sqrt{2}$.

$$\frac{1\times\sqrt{2}}{\sqrt{2}\times\sqrt{2}}=\frac{1\times\sqrt{2}}{2}=\frac{\sqrt{2}}{2}$$

See? The denominator is now an integer, and everything's all nice and legal.

Roots Quick Quiz

Question 1 of 4

$5\sqrt{18} \times 2\sqrt{10} =$

- ○ $20\sqrt{7}$
- ○ $30\sqrt{2}$
- ○ $40\sqrt{3}$
- ○ $50\sqrt{6}$
- ○ $60\sqrt{5}$

Question 2 of 4

$12\sqrt{60} \div 2\sqrt{5} =$

- ○ $6\sqrt{55}$
- ○ $12\sqrt{3}$
- ○ $18\sqrt{2}$
- ○ $60\sqrt{5}$
- ○ $240\sqrt{3}$

Question 3 of 4

Quantity A	**Quantity B**
The number of integer multiples of 3 between $\sqrt{100}$ and $\sqrt{1000}$	21

- ○ Quantity A is greater.
- ○ Quantity B is greater.
- ○ The two quantities are equal.
- ○ The relationship cannot be determined from the information given.

Question 4 of 4

$3\sqrt{18}+2\sqrt{50}=$

○ $10\sqrt{17}$

○ $12\sqrt{17}$

○ 15

○ $19\sqrt{2}$

○ 180

Explanations for Roots Quick Quiz

1. Multiply the numbers outside the square-root sign first: $5 \times 2 = 10$. Next, multiply the radicands: $18 \times 10 = 180$. The combined expression is $10\sqrt{180}$, but you're not done. The largest perfect square that is a factor of 180 is 36, so factor it out: $\sqrt{180}=\sqrt{36\times 5}=\sqrt{36}\times\sqrt{5}=6\sqrt{5}$. The new expression is $10 \times 6\sqrt{5}$, which combines to $60\sqrt{5}$. The answer is (E).

2. Divide the coefficients first: $12 \div 2 = 6$. Now you can divide the radicands to get a new one: $\sqrt{60}\div\sqrt{5}=\sqrt{\frac{60}{5}}=\sqrt{12}$, which, when simplified, becomes $2\sqrt{3}$. The new expression is $6 \times 2\sqrt{3}$, which combines to $12\sqrt{3}$. The answer is (B).

3. The $\sqrt{100}$ is 10. Use a calculator to find that $\sqrt{1000}$ is a number between 31 and 32. Now count the multiples of 3: 12, 15, 18, 21, 24, 27, and 30. There are six. The answer is (B). Alternatively, you could start at 12, the first multiple of three greater than 10, and count up 21 multiples of three. That gets you to 72. 72^2 is greater than 5,000, so clearly 21 multiples is far too many.

4. You can't do a thing until you convert the terms so that they have the same radicand. The first term, $3\sqrt{18}$, can be simplified to $3 \times 3\sqrt{2}$, or $9\sqrt{2}$. The second term, $2\sqrt{50}$, can be rewritten as $2 \times 5\sqrt{2}$, or $10\sqrt{2}$. Now, you can add the terms, which combine to $19\sqrt{2}$. The answer is (D).

Whew. This was a long chapter, chock full of gooey, mathematical goodness. To see how well you retained it, try these questions and review the answers that immediately follow. We'll see you over in the next chapter, which tells you how much you should know about algebra—and how little of it you should use.

Nuts and Bolts Drill

Let's review and test your new skills on the nuts and bolts of GRE math in the following drill. Remember to work carefully!

Question 1 of 20

What is the sum of the distinct prime factors of 36 ?

- ○ 2
- ○ 5
- ○ 6
- ○ 8
- ○ 10

Question 2 of 20

If a certain fraction with a numerator of 12 has a value of 0.25, then the denominator is

- ○ 3
- ○ 12
- ○ 25
- ○ 48
- ○ 300

Question 3 of 20

If 5 fewer than $\frac{40}{x}$ is –1, then x =

- ○ –10
- ○ –4
- ○ 4
- ○ 8
- ○ 10

Question 4 of 20

The numbers that correspond to points A, B, and C on the number line are $-\frac{5}{3}$, $-\frac{2}{3}$, and, $\frac{2}{3}$ respectively. Which of the following values fall between the average of A and B and the average of B and C ?

- ☐ –1
- ☐ $-\frac{2}{3}$
- ☐ 0
- ☐ $\frac{1}{2}$
- ☐ 1

Question 5 of 20

If $\frac{0.0017}{x} > 1$, then each of the following could be the value of x EXCEPT

- ○ 0.0011
- ○ 0.0013
- ○ 0.0015
- ○ 0.0017
- ○ 0.0091

Question 6 of 20

$27^0 + 36^0 =$

- ○ 0
- ○ 2
- ○ 9
- ○ 15
- ○ 63

Question 7 of 20

How many positive integers that are multiples of 3 are also divisors of 42 ?

- ○ One
- ○ Two
- ○ Three
- ○ Four
- ○ Five

Question 8 of 20

In a certain hardware store, 3 percent of the lawnmowers need new labels. If the price per label is $4 and the total cost for new lawnmower labels is $96, how many lawnmowers are in the hardware store?

- ○ 1,600
- ○ 800
- ○ 240
- ○ 120
- ○ 24

Question 9 of 20

Quantity A	**Quantity B**
$15 - 18 \div (7 - 4)^2 \times 8 + 2$	1

- ○ Quantity A is greater.
- ○ Quantity B is greater.
- ○ The two quantities are equal.
- ○ The relationship cannot be determined from the information given.

Question 10 of 20

What is the greatest possible value of integer n if $6^n < 36^{10}$?

- ○ 6
- ○ 9
- ○ 12
- ○ 15
- ○ 19

Question 11 of 20

5×10^3 is what percent of $\frac{1}{5} \times 10^2$?

- ○ 2,500%
- ○ 4,900%
- ○ 20,000%
- ○ 24,900%
- ○ 25,000%

Question 12 of 20

50 percent of 2000 ÷ 75 percent of

$$400 - \frac{(0.3)(0.2)(0.5)}{(0.04)(0.09)(0.25)} =$$

- ○ –30
- ○ 0
- ○ 9
- ○ 30
- ○ 90

Question 13 of 20

Quantity A	**Quantity B**
$\dfrac{9^{17} - 9^{15}}{12^2}$	$9^{14} \times 5$

- ○ Quantity A is greater.
- ○ Quantity B is greater.
- ○ The two quantities are equal.
- ○ The relationship cannot be determined from the information given.

Question 14 of 20

A is a positive odd number less than 5. Which of the following could be the value of $\dfrac{(10 + A)}{2} + 3$?

- ○ 9
- ○ 9.5
- ○ 10
- ○ 10.5
- ○ 11

Question 15 of 20

In the number 4,*A*34, the letter *A* represents a digit. For which of the following values of *A* is 4,*A*34 a multiple of 3 ?

- ○ 2
- ○ 3
- ○ 4
- ○ 6
- ○ 9

Question 16 of 20

At a certain animal shelter, the annual budget for cat food was $3,900 ten years ago. The current cat food budget is 125 percent greater than the budget was ten years ago. What is the amount, in dollars, of the current annual cat food budget?

[ ]

Question 17 of 20

Which of the following are true statements?

I. $5^3 \times 5^{\frac{1}{3}} = 1$

II. $5 \times \sqrt{5^{-2}} = 1$

III. $\sqrt{5} \times \dfrac{\sqrt{5}}{5} = 1$

- ○ I only
- ○ II only
- ○ III only
- ○ I and II only
- ○ II and III only

Question 18 of 20

Quantity A	Quantity B
$\sqrt{16+121}$	15

- ○ Quantity A is greater.
- ○ Quantity B is greater.
- ○ The two quantities are equal.
- ○ The relationship cannot be determined from the information given.

Question 19 of 20

If 15% of j is greater than 55% of k, and $325 < k < 375$, then which of the following are possible values for j ?

- ☐ 57
- ☐ 332
- ☐ 782
- ☐ 1,087
- ☐ 1,192
- ☐ 1,314

Question 20 of 20

How many more integers, between 552 and 652, exclusive, are divisible by 3 than are divisible by 4 ?

- ○ 1
- ○ 6
- ○ 8
- ○ 9
- ○ 12

EXPLANATIONS FOR NUTS AND BOLTS DRILL

1. B

Find the prime factors of 36: 2, 2, 3, and 3. *Distinct* means *different,* so the distinct prime factors are 2 and 3, and the sum is 5.

2. D

0.25 is equivalent to $\frac{1}{4}$. If you multiply the top and bottom of the fraction by 12, you get $\frac{12}{48}$. The denominator must be 48.

3. E

Translate "5 fewer than $\frac{40}{x}$ is –1" into an equation: $\frac{40}{x} - 5 = -1$, or $\frac{40}{x} = 4$. Solve for x or look for the answer choice that works. The answer is choice (E) because $\frac{40}{10} = 4$.

4. A and B

The average of A and B is $(-\frac{5}{3} + -\frac{2}{3}) \div 2 = -\frac{7}{6}$. The average of B and C, using the same type of calculation, equals 0. Only choices (A) and (B) fall within this range.

5. E

Use your on-screen calculator, or write out each answer choice as the denominator of a fraction with 0.0017 as the numerator. Move the decimal points to the right the same number of places on both top and bottom until it's clear whether the fraction is greater than or equal to 1. Either way, be careful about choice (D), which equals 1, and so isn't correct. Only choice (E) is not greater than or equal to 1: $\frac{0.0017}{0.0091} = \frac{1.7}{9.1}$, which is less than 1.

6. B

By definition, any number raised to the power of 0 is equal to 1, and 1 + 1 = 2.

7. D

You're looking for numbers that are both multiples of 3 and divisors of 42. There is an infinite number of multiples of 3 but a limited number of divisors of 42, so start there. The divisors (factors) of 42 are 1, 2, 3, 6, 7, 14, 21, and 42. Which of these are also multiples of 3? There are four numbers that fit the bill—3, 6, 21, and 42, so choice (D).

8. B

Let's let x be equal to the number of lawnmowers in the hardware store. Then $.03x$ will be equal to the number of lawnmowers in need of new labels, and $4(.03x)$ will be equal to the total cost of these lawnmowers in need of new labels. But we know the total cost: it's \$96. So, we can set $4(.03x) = 96$ and solve for x. 96 divided by 4(.03) is 800, or (B).

9. C

The order of operations is the key here: Parentheses, Exponents, Multiplication and Division, Addition and Subtraction. Quantity A becomes 15 – 18 ÷ 9 × 8 + 2. The division and multiplication go in left to right order, so 15 – (18 ÷ 9) × 8 + 2 = 15 – (2 × 8) + 2 = 15 – 16 + 2 = 1. The two quantities are always equal and choice (C) is correct.

10. E

36 is really 6^2 in disguise. So $= 36^{10} = \left(6^2\right)^{10}$ or 6^{20}. 19 is the greatest possible value for n that is still smaller than 6^{20}. The answer is (E).

11. E

Translate into an equation: $5 \times 10^3 = \frac{x}{100} \times \frac{1}{5} \times 10^2$. $\frac{5 \times 10^3}{10^2} = \frac{x}{100} \times \frac{1}{5}$, or $50 = \frac{x}{500}$. Therefore, $x = 25{,}000$.

12. A

Follow the order of operations. First, convert the percents to decimals and rewrite the problem using mathematical symbols: $(0.5 \times 2000) \div (0.75 \times 400) - \left(\frac{(0.3)(0.2)(0.5)}{(0.04)(0.09)(0.25)}\right)$. Next, solve both parentheses before the subtraction sign: $1000 \div 300 = \frac{10}{3}$. Now convert the decimals to integers in the second fraction by moving the decimal point two places to the right in each term of the numerator and denominator, and reduce: $\frac{30 \times 20 \times 50}{4 \times 9 \times 25} = \frac{100}{3}$. Finally, subtract the two fractions: $\frac{10}{3} - \frac{100}{3} = -\frac{90}{3} = -30$

13. C

If a problem asks you to add or subtract large exponents, it is an opportunity to factor. You can factor out 9^{15} from the top of the fraction in Quantity A, giving you 9^{15} $(9^2 - 1)$ which equals 9^{15} (80). The next step is to see if you can factor out the 12^2 in the denominator. 12 is a multiple of both 3 and 4, so 12^2 will have 3^2 and 4^2. Since $3^2 = 9$, you can subtract one 9 from the numerator; $4^2 = 16$, which does not contain any threes or nines to cancel. But 80 ÷ 16 = 5, so after simplifying your fraction you get $9^{14} \times 5$. Since this matches the value in Quantity B, the correct answer is choice (C).

14. B

Set up the equation so that $\frac{10}{2} + \frac{a}{2} + \frac{6}{2} =$ one of the answer choices. This can be simplified to $\frac{16}{2} + \frac{a}{2} =$ an answer choice. If you convert each of the answer choices to an improper fraction with 2 as the denominator, the math becomes quite simple. For answer choice A you get: $\frac{16}{2} + \frac{a}{2} = \frac{18}{2}$, so A = 2. Cross off (A). For (B), A = 3 for (C), A = 4, for (D), A = 5, for (E), A = 6. Only answer choice (B) works.

15. C

Test the answer choices using the divisibility-of-three rule: The sum of the digits should be divisible by three. For choice (A), the number is 4,134. The sum of the digits is 4 + 2 + 3 + 4 = 13. Because 13 is not divisible by 3, 4,134 is not divisible by three. For choice (B), the sum is 4 + 3 + 3 + 4 = 14, which is not divisible by three. For the choice (C), the sum is 4 + 4 + 3 + 4 = 15, which is divisible by three. For choice (D), the sum is 4 + 6 + 3 + 4 = 17, which is not divisible by three. For choice (E), the sum is 4 + 9 + 3 + 4 = 20, which is not divisible by three. The answer, therefore, is (C).

16. 8,775

Since the question says *125 percent greater than*, use the percent change formula: Percent change is $\left(\frac{\textit{difference}}{\textit{original}}\right)\times 100$. This means that $125\% = \left(\frac{\textit{difference}}{3{,}900}\right)$ and you can solve this to get \$4,875. This is the difference, not the total for the current year. To get the current total, add \$4,875 to the original \$3,900 to get the correct amount 8,775.

17. E

Evaluate each equation. The first one equals5 $5^{\frac{4}{3}}$, which does not equal 1, so eliminate choices (A) and (D). The next equation equals $5 \times \sqrt{\frac{1}{25}}$, or $5 \times \frac{1}{5}$; this equation works, so eliminate choice (C). The third equation equals $\frac{\sqrt{25}}{5}$, which also equals 1, so this equation works as well; choice (E) is correct.

18. B

Simplify the value in Quantity A to make it easier to approximate: $\sqrt{16+121} = \sqrt{137}$. Since $\sqrt{144} = 12$, the value in Quantity A is less than 12, and the value in Quantity B is greater. Alternately, you could turn the value in Quantity B into a root: $15 = \sqrt{225}$, and $\sqrt{225}$ is greater than $\sqrt{137}$.

19. E and F

Use percent translation to establish the smallest possible value of *j*: $\frac{15}{100} \times j > \frac{55}{100} \times 325$. Solve the inequality for *j* to get $j > 1191\frac{2}{3}$. Since no limits are set on the greatest possible value of *j*, any value greater than $1191\frac{2}{3}$ will work.

20. D

The most reliable way to approach this question is to count out the multiples of 3 and of 4, then subtract. It appears you have a range of 100, which when divided by 3 equals approximately 33, and when divided by 4 equals 25. 33 – 25 = 8, which is choice (C), but this is incorrect. There are 33 multiples of 3, but since 552 and 653 are excluded, there are only 24 multiples of 4. 33 – 24 = 9, which is choice (D).

Chapter 4
Algebra, and How to Get Rid of It

ALGEBRA

If you're planning to attend graduate school, you've probably had some sort of algebraic training in the dark reaches of your past. Algebra is the fine art of determining how variable quantities relate to each other within complex functions, and it dates back more than 3,000 years to ancient Babylon.

If it seems like it's been 3,000 years since you last studied algebra, or even if you flunked algebra last year, fear not. This chapter is devoted to the following two very important pursuits:

- refreshing your memory of the basic algebraic rules that the new GRE tests, and
- removing almost all algebra from your GRE experience.

It's impossible—and *not* in your best interests—to ignore algebra entirely on the GRE, because there will likely be a few questions that are best solved by using algebra. But for the most part, the best strategy is to embrace new techniques that help you look at these problems from a different perspective. Naturally, the best way to subvert the rules is to review them first.

Your Mileage May Vary

Any letter in an algebraic term or equation is called a *variable*; you don't know what its numerical value is. It varies. Until you solve an equation, a variable is an unknown quantity. Any number that's directly in front of a variable is called a *coefficient*, and the coefficient is a constant multiplied by that variable. For example, $3x$ means "three times x," whatever x is.

Combining Like Terms

If two terms have the same variables or series of variables in them, they're referred to as *like terms*. You can combine them like this:

$$6a + 4a = 10a$$

$$13x - 7x = 6x$$

For example, if you have six apples in one hand and four in the other, you have a total of 10 apples (and a pair of humongous hands).

Solving an Equation

Whenever a variable appears in an equation and you have to find the value of the variable, you have to "isolate" it by following these steps.

- **Step One:** Use addition and/or subtraction to put all the terms that contain the variable on one side of the equal sign and all terms that don't contain the variable on the other.
- **Step Two:** Use multiplication or division to remove whatever coefficient the variable has, until the variable is sitting there all by itself.

In order to solve an equation, you must rely on the following paramount rule of algebraic manipulation:

> You can do anything you want to an equation as long as you do exactly the same thing to both sides.

Question 5 of 20

If $4x - 5 = 19$, what is the value of x ?

[ ]

Here's How to Crack It

In order to get all variable terms on one side of the equal sign and all constant terms on the other, follow Step One and add 5 to both sides.

$$4x - 5 + 5 = 19 + 5$$

The 5's on the left cancel out, and the equation becomes $4x = 24$. There's no more addition or subtraction to be done, so you're through with Step One.

All that's left is to deal with the coefficient: 4. Because $4x$ equals "4 times x," you can simplify the equation by dividing both sides by 4:

$$\frac{4x}{4} = \frac{24}{4}$$
$$x = 6$$

Always Check

Taking the GRE involves a lot of stress and fatigue, so it's easy to make a little boo-boo when you're manipulating an algebraic equation. Therefore, it always pays to plug your answer back into the original problem to make sure it works. Let's do it.

$$4(6) - 5 = 19$$

$$24 - 5 = 19$$

$$19 = 19. \text{ Check.}$$

Keep in mind that solutions to equations don't always have to be integers, so don't be concerned if your result is a fraction. As long as it works when you plug it back into the equation, you're fine.

Question 4 of 25

If $4m + 7 = 16 - 2m$, what is the value of m ?

[ ]

Here's How to Crack It

Step One

Get all the variables onto the left side of the equation by adding $2m$ to both sides, then move all the constants to the right side by subtracting 7 from both sides.

$$4m + 7 + 2m = 16 - 2m + 2m$$

$$6m + 7 = 16$$

$$6m + 7 - 7 = 16 - 7$$

$$6m = 9$$

Step Two

Divide both sides by 6.

$$\frac{6m}{6} = \frac{9}{6}$$

$$m = \frac{9}{6} = \frac{3}{2}$$

Step Three
Check your work by plugging in.

$$4\left(\frac{3}{2}\right)+7=16-2\left(\frac{3}{2}\right)$$
$$\frac{12}{2}+7=16-\frac{6}{2}$$
$$6+7=16-3$$
$$13 = 13. \text{ Check.}$$

Take special note of Step 3 now and be prepared to use this skill a lot in the latter half of this chapter.

Solving an Equation Quick Quiz

Solve for x in each of the following:

1. $2x - 5 = 11$
2. $3 - 5x = 13$
3. $12x + 4 = 4x$
4. $8x - 9 = x - 2$
5. $\frac{x}{2} + 5 = -9$

Explanations for Solving an Equation Quick Quiz

1. $x = 8$
2. $x = -2$
3. $x = -\frac{1}{2}$
4. $x = 1$
5. $x = -28$

Inequalities

Inequality symbols are used to convey that one number is greater than or less than another.

> The symbols used in inequalities are as follows:
>
> $>$ means "is greater than"
> $<$ means "is less than"
> $\geq$ means "is greater than or equal to"
> $\leq$ means "is less than or equal to"

Even though the two sides of an inequality aren't equal, you can treat them in much the same way as you do the expressions in regular equations, when you have to solve for a variable.

Question 13 of 20

If $5b - 3 > 2b + 9$, which of the following must be true?

- ○ $b > -4$
- ○ $b > 2$
- ○ $b < 2$
- ○ $b > 4$
- ○ $b < 4$

Here's How to Crack It

Adding and subtracting take place as usual, like this:

$$5b - 3 - 2b > 2b + 9 - 2b$$

$$3b - 3 > 9$$

$$3b - 3 + 3 > 9 + 3$$

$$3b > 12$$

At this point, because the coefficient of b is positive, you can divide both sides by 3 and get a final range of values for b.

$$\frac{3b}{3} > \frac{12}{3}$$
$$b > 4$$

To check your solution, try a number that is greater than 4 (say, 5) and see if the inequality holds true.

$$5(5) - 3 > 2(5) + 9$$

$$24 - 3 > 10 + 9$$

$21 > 19$. Check. The answer is (D).

Flip That Sign!

The only difference between solving equalities and inequalities is this one very important rule:

> Whenever you multiply or divide both sides of an inequality by a negative number, you must flip the inequality sign.

Let's try a problem.

Question 16 of 20

If $5 - 11p \leq 9$, which of the following could be the value of p ?

*Indicate **all** possible values.*

- ☐ −15
- ☐ −10
- ☐ −2
- ☐ 0
- ☐ 1
- ☐ 8
- ☐ 24

Here's How to Crack It

Hey, look. It's another one of those newfangled multi-answer questions. How exciting. Truly.

Manipulate the problem as you would a regular equality, like so:

$$5 - 11p \leq 9$$

$$5 - 11p - 5 \leq 9 - 5$$

$$-11p \leq 4$$

Now that you're dividing by –11, flip the sign and you're done.

$$\frac{-11p}{-11} \leq \frac{4}{-11}$$

$$p \geq -\frac{4}{11}$$

Any value that is greater than $-\frac{4}{11}$ is a possible value of p, so you should check the box next to 0 and every other box with a number that's greater than zero. See how important that little rule is? If you didn't know about it, you might have picked all the numbers that were less than $-\frac{4}{11}$, which would be the exact opposite of the correct answers. And that would have been unfortunate.

Everybody Dance

Pretend you had two guys, Al and Bob, who wanted to each dance with two gals: Cathy and Daisy. To make sure each guy danced with each gal, we could first have Al dance with Cathy, and then have Al dance with Daisy. Now that Al's danced with both ladies, Bob gets to dance with Cathy, and then Bob gets to dance with Daisy. Because we systematically paired Al up with each possibility, and then paired Bob up with each possibility, we've set up every possible pair of dance partners.

Why is this important? Some inequalities questions will ask you to find the range for two combined inequalities, and we'll answer these questions similarly. For that type of question, you will need to find every possible result of combining the two inequalities: The lowest number of the first inequality will have to be paired up with the lowest and highest numbers of the second inequality, and the highest number of the first inequality will be paired up with the lowest and highest numbers of the second inequality. We'll end up with a total of four numbers: Pick the lowest and highest, and that's your combined range.

Question 12 of 20

If $-5 \le a \le 12$ and $-10 \le b \le 25$, which of the following represents all possible values of $a - b$?

- ○ $-30 \le a - b \le 22$
- ○ $-30 \le a - b \le 5$
- ○ $-13 \le a - b \le 5$
- ○ $5 \le a - b \le 22$
- ○ $5 \le a - b \le 30$

Here's How to Crack It

For the range of *a*, we have two numbers: –5 and 12. We'll need to make sure that each of those numbers gets to dance with each of the two numbers from the range of *b*.

$$a - b$$
$$(-5) - (-10) = 5$$
$$(-5) - (25) = -30$$
$$(12) - (-10) = 22$$
$$(12) - (25) = -13$$

Notice how each number from *a* was paired up with each number from *b*? Our smallest result for $a - b$ was –30, and our highest was 22, so the full range is $-30 \le a - b \le 22$, answer (A).

Inequalities Quick Quiz

In questions 1-3, find the range of values of x.

1. $3x + 7 > 22$
2. $8 - 3x < 13$
3. $-\frac{x}{2} + 2 \ge 5x - 9$

Question 4 of 6

If y is a positive integer and $6y + 9 > 5 + 8y$, then $y =$

Question 5 of 6

If $5f + 11 \geq 17 + f$ and $7 - 4f > -13$, f could equal each of the following EXCEPT:

- ○ 1.6
- ○ 2.4
- ○ 3.9
- ○ 4.1
- ○ 5.3

Question 6 of 6

If $-7 \leq x \leq 5$ and $-15 \leq y \leq 0$, what is the range of all possible values of xy ?

Explanations for Inequalities Quick Quiz

1. $x > 5$
2. $x > -\frac{5}{3}$
3. $x \leq 2$
4. If you solve the inequality, you find that $y < 2$. The only positive integer less than 2 is 1, so x must equal 1.
5. Solve both inequalities. If $5f + 11 \geq 17 + f$, then $f \geq 0.5$, and if $7 - 4f > -13$, then $f < 5$. Therefore, f could be anywhere in the range of 0.5 to 5. The only answer choice that is not in this range is (E).
6. Let's pair up our biggest and smallest from each inequality:

   $x \times y$

   $(-7) \times (-15) = 105$

   $(-7) \times (0) = 0$

   $(5) \times (-15) = -75$

   $(5) \times (0) = 0$

   So the range is $-75 \leq xy \leq 105$

Quadratics

Remember all that factoring and FOILing you did in high school algebra? Well, you probably won't have to go through all the trial and error of factoring, but there will be circumstances in which you'll have to combine a pair of binomials. (That's just a "mathism" for an algebraic element that contains two terms, like "$2x + y$.") For that, make like a fencer and whip out your FOIL.

FOIL

When FOILing you combine two binomials by multiplying the First, Outside, Inside, and Last terms, and then simplify wherever possible. Let's give it a shot.

Question 11 of 20

Find the product of $(x - 3)$ and $(2x + 7)$.

Here's How to Crack It

If you're new to FOILing, it helps to line up your products so you can keep track of what you're doing:

Firsts: $x \cdot 2x = 2x^2$

Outsides: $x \cdot 7 = 7x$

Insides: $-3 \cdot 2x = -6x$

Lasts: $-3 \cdot 7 = -21$.

Now combine: $2x^2 + 7x - 6x - 21 = 2x^2 + x - 21$.

Solving Quadratic Equations

When quadratics and equal signs come together, the result is a quadratic equation. On the GRE, quadratics are usually set equal to zero, and you'll have find an equation's solutions, or roots, by factoring.

Question 2 of 20

What positive value of x satisfies the equation $x^2 - 5x - 6 = 0$?

Here's How to Crack It

To answer this question, you need to solve this equation (or, in other words, find the equation's roots). In other words, you'll have to factor the quadratic. Factoring is basically the opposite of FOILing, and it usually requires a little trial and error. Placing the two *x*'s in the parentheses is the easy part. The challenge lies in finding two numbers whose sum is –5 (the middle coefficient) and whose product is –6 (the last term). In this case, those numbers are –6 and 1.

$$x^2 - 5x - 6 = 0$$

$$(x \quad\;)(x \quad\;) = 0$$

$$(x - 6)(x + 1) = 0$$

In order for the product of two numbers to be 0, one of them must be 0. So set both of your factors equal to 0 and solve for x.

$$x - 6 = 0 \qquad\qquad x + 1 = 0$$

$$x = 6 \qquad\qquad x = -1$$

Check your answers to make sure you didn't make any unfortunate slip-ups.

$$6^2 - 5(6) - 6 = 0 \qquad\qquad (-1)^2 - 5(-1) - 6 = 0$$

$$36 - 30 - 6 = 0 \qquad\qquad 1 + 5 - 6 = 0$$

$0 = 0$. Check. $\qquad\qquad 0 = 0$. Check.

Now, we know 6 and –1 are the roots of the equation, but the question only asks for the positive root. So the answer is 6.

Common Quadratics

As basic as FOIL is, there are a few very common multiplications of binomials that come up so frequently that you're better off memorizing them. Standardized-test writers like them a lot, because they're great for making easier problems seem a lot more difficult.

$$(x + y)^2 = x^2 + 2xy + y^2$$

$$(x - y)^2 = x^2 - 2xy + y^2$$

$$(x + y)(x - y) = x^2 - y^2$$

Knowledge of the last formula—which is commonly referred to as a "difference of squares"—is especially useful if you come across a question that looks like this:

Question 14 of 20

If $\frac{x^2 - 9}{x - 3} = 7$, what is the value of x ?

Here's How to Crack It

Rather than resort to cross-multiplication, here you can recognize that the fraction on the left is in the form of $\frac{x^2 - y^2}{x - y}$; in this case, y is a constant. Because $\frac{x^2 - y^2}{x - y} = \frac{(x + y)(x - y)}{(x - y)} = x + y$, you can rewrite the left side as $\frac{x^2 - 9}{x - 3} = x + 3$, and your math becomes easy. If $x + 3 = 7$, then $x = 4$.

Here's another one on which a little quadratic knowledge can save you some time.

Question 19 of 20

Quantity A	Quantity B
$\frac{836^2 - 835^2}{836 + 835}$	1

- ○ Quantity A is greater.
- ○ Quantity B is greater.
- ○ The two quantities are equal.
- ○ The relationship cannot be determined from the information given.

Here's How to Crack It

If you look closely, you can see that the given term follows the format $\frac{x^2 - y^2}{x - y} = x + y$, so Quantity A must be 836 – 835, or 1. Therefore, the answer is (C). You could also use your calculator.

Quadratics Quick Quiz

1. If $x^2 - 5x + 6 = 0$, what is a possible value of x ?
2. If $4x^2 + 7x - 2 = 0$, what is a possible value of x ?
3. If $8x^2 + 14x - 15 = 0$, what is a possible value of x ?
4. $\frac{x^2 - 169}{x - 13} = 15$, then $x =$
5. If $a + b = 5$ and $a^2 + b^2 = 15$, then $ab =$

Explanations for Quadratics Quick Quiz

1. Because $x^2 - 5x + 6$ can be factored to $(x - 2)(x - 3)$, then the possible values of x are 2 and 3.

2. Because $4x^2 + 7x - 2$ can be factored to $(4x - 1)(x + 2)$, the two possible values of x are 0.25 and –2.
3. This one's a little trickier to factor, and it takes a lot more trial and error, but the factors of $8x^2 + 14x - 15$ are $(4x - 3)$ and $(2x + 5)$. Therefore, the two possible values of x are $\frac{3}{4}$ and $-\frac{5}{2}$.
4. Because $\frac{x^2 - 169}{x - 13} = \frac{(x+13)(x-13)}{(x-13)}$, the equation can be rewritten as $x + 13 = 15$. Therefore, $x = 2$.
5. This question makes you think you have to solve for a and b, but you don't. If $a + b = 5$, then $(a + b)^2 = 5^2$, or 25. When you expand the term on the left, $(a + b)^2$ becomes $a^2 + 2ab + b^2$. Rewrite the equation as $a^2 + b^2 + 2ab = 25$; because $a^2 + b^2 = 15$, you can substitute 15 in the equation like this: $15 + 2ab = 25$. Therefore, $2ab = 10$, and $ab = 5$.

Simultaneous Equations

If a single equation has two variables, you can't solve for either one. If $2x + y = 5$, for example, then $x = 2$ and $y = 1$ could be one solution; but $x = 1$ and $y = 3$ could be another. But if you have two distinct equations and two variables—often referred to in high school math courses as "two equations, two unknowns"—then each variable has only one possible solution.

If a question asks you to find the value of one variable, you can usually add or subtract the equations and solve.

Question 16 of 20

If $2x - 3y = 7$ and $x + 3y = 8$, what is the value of x ?

Here's How to Crack It

Look at the coefficients of the y terms; they're opposites of each other. In other words, if you add them together, you get zero. So line up the equations like this and add all the like terms separately.

$$\begin{array}{r} 2x - 3y = 7 \\ \underline{x + 3y = 8} \\ 3x + 0 = 15 \end{array}$$

From here, you can determine that $x = 5$.

When There's Less Work Than You Think

ETS likes to build its simultaneous equation questions to look a lot more daunting and time-consuming than they actually are. Take this equation for example.

Question 9 of 20

If $a + 3b = 10$ and $a - b = 8$, what is the value of $a + b$?

At first glance, you might think you have to solve for a and b individually and then add them together to get your final answer. But that isn't the case.

Here's How to Crack It

If you add the two equations together, you get a new equation.

$$\begin{array}{r} a + 3b = 10 \\ \underline{a - b = 8} \\ 2a + 2b = 18 \end{array}$$

If you divide each term in this equation by 2, you will see the answer right in front of you.

$$2(a + b) = 18$$

$$\frac{2(a+b)}{2} = \frac{18}{2}$$

$$a + b = 9$$

As it turns out, we don't have to care what the individual values of x and y are. All we have to know is that their sum is 9.

Simultaneous Equations Quick Quiz

1. If $2x - 3y = 2$ and $x - 5y = -6$, what is the value of y ?
2. If $3a - b = 200$ and $5a + 2b = 40$, what is the value of $a - b$?

3. If $4m - 3n = 12$ and $m - 2n = 3$, then $m - n =$

4. At a stationery store, three pens and five notebooks cost a total of $19.02 and two pens and three notebooks cost a total of $11.63. What is the price of one notebook?

Answers to Simultaneous Equations Quick Quiz

1. Because you're solving for y, you want to make the x's disappear. You can do this by multiplying the second equation by –2 so that it becomes $-2x + 10y = 12$. Now, add the equations and the x's drop out.

$$2x - 3y = 2$$

$$-2x + 10y = 12$$

$$7y = 14$$

$$y = 2$$

2. Solve for a first by multiplying the first equation by 2: $6a - 2b = 400$. Now add.

$$6a - 2b = 400$$

$$5a + 2b = 40$$

$$11a = 440$$

$$a = 40$$

Now plug 40 into either equation to find b: $3(40) - b = 200$, so $b = -80$. Therefore, $a - b = 40 - (-80) = 120$.

3. This one makes you think you have to solve for both variables individually, but you don't if you just stack them and add them.

$$4m - 3n = 12$$

$$m - 2n = 3$$

$$5m - 5n = 15$$

$$5(m - n) = 15$$

$$m - n = 3$$

4. This is a simultaneous equation problem masquerading as a word problem. If three pens and five notebooks cost a total of \$19.02, then $3p + 5n = 19.02$. Similarly, the second equation can be rewritten as $2p + 3n = 11.63$. To find the value of n, get rid of the p's.

$$2(3p + 5n = 19.02)$$

$$-3(2p + 3n = 11.63)$$

$$6p + 10n = 38.04$$

$$-6p - 9n = -34.89$$

$$n = 3.15$$

Each notebook costs \$3.15.

All right, that was a lot of algebra to learn, especially after we said that you should use as little algebra as possible on test day. But now we'll talk about *why* you should use as little algebra as possible.

WHY ALGEBRA IS NOT TO BE TRUSTED

What is x, anyway? What does it mean to have y marbles in a jar? Or z chairs at a table? Nothing, that's what. Algebra has been the backbone of mathematics for millennia, but that doesn't mean it will do you any good on the GRE. In fact, it's more likely to trip you up.

Let Us Count the Ways

A basic appreciation of algebra is crucial for a good quantitative score on the GRE. But whenever you have the option, you should use arithmetic instead of algebra, for a number of reasons.

- Algebra is based on abstract unknowns, while arithmetic is a concrete system of numbers you can visualize (x apples versus 2 apples).
- You started doing arithmetic long before you even knew what algebra was, so arithmetic is far more ingrained in your brain.
- You probably haven't done a lick of algebra since you finished school, but you perform some sort of arithmetic every day (whether you realize it or not), so arithmetic is far more familiar.
- Your calculator can help you with arithmetical calculations, but it's helpless on algebraic manipulations.

Some of you out there might think your algebra skills are passable, and that you're not worried about making mistakes. That may well be, but we're here to tell you that things change once you've left the friendly confines of preparing and you're actually taking the GRE for real. Stress happens, and even the most brilliant math students can choke under pressure.

The other bad thing about algebra is that often when you're performing it, you can mess something up and not even know it—because the answer you get, the wrong answer—is usually right there among the answer choices.

Do the Unexpected

If there were two roads between your house and the center of town and you knew one of them was full of landmines, which road would you choose?

This brings us to the most important reason not to use algebra. Test writers know that most students have been trained to use algebra when they see an algebraic question, and that many students make lots of careless mistakes. Therefore, most of the wrong answers test writers dream up to use in questions are based on the errors they anticipate you'll make. If you avoid using algebra as much as possible, you'll avoid many of the pitfalls that ETS has laid out for you.

So when you see a problem with a bunch of variables in it, don't think the way they think you'll think. Instead, plug in numbers.

PLUGGING IN

Plugging In is a lot like writing your own novel; rather than wonder how many candies Phil has in his hand, assume the power to write the narrative. Decide *for yourself* how many he has.

When to Do It

There are two dead giveaways when you're identifying problems that you can solve using your own numbers.

- The problems feature the phrase "in terms of."
- The answer choices have variables in them.

1. **Recognize the Opportunity to Plug In.** If you have variables in the answer choices, plug in. If there is an unknown quantity in a problem that cannot be solved for, plug in.
2. **Set Up Your Scratch Paper.** Write down letters for each answer choice and any important information from the problem.
3. **Plug In an Easy Number for the Variable.** Choose an easy number, such as 2, 3, 5, 10, or 100, for one of the variables. Write clearly what number you're plugging in for which variable. If the problem gives you certain limitations for the value of the variable, such as "odd integer larger than 30," be sure to choose a number that follows those limitations. If there are multiple variables in a problem, see if you can solve for the other variables once you've plugged in for the first one. If you can't, plug in for those variables as well.
4. **Find Your Target Number.** Solve the question using your numbers. Whatever is asked at the end of the question is your target number. Write it on your scratch paper and circle it.
5. **Check All of the Answer Choices.** Using the numbers that you plugged into the question, try out every answer choice. Put a check mark next to any answer that gives you your target number, but keep checking. You must check every answer. Most of the time, only one answer choice will work out, but if you end up with two or more answer choices that give you your target number, plug in a new set of numbers, find a new target number, and check the remaining answer choices.

Let's try an example:

Trigger: Variables in the answer choices.

Response: Plug in.

Question 11 of 20

George is twice as old as Mary, and Mary is three years older than Juan. If Juan is j years old, then what, in terms of j, will be George's age in 10 years?

- ○ $2j - 4$
- ○ $2j - 14$
- ○ $2j + 13$
- ○ $2j + 16$
- ○ $2j + 26$

Here's How to Crack It

Look at the answer choices first. You should **recognize the opportunity to plug in** as soon as you see answer choices like those. There are variables in all of our answer choices, which means this is a perfect plug in problem. Now that we know what we're going to do, don't sit and stare at the problem. We're going to need to **set up our scratch paper** by writing down A B C D E. Since our variable is j, let's **plug in an easy number** for Juan's age. Let's say Juan is 5, so $j = 5$.

Now we can work through the problem. Take it apart piece by piece. The first part of the problem states that George is twice as old as Mary, which is nice for George and Mary, but we know only how old Juan is. Let's leave that part of the problem alone, and move on. The next part of the sentence states that Mary is three years older than Juan. Hey, now there's something we can deal with. Since Juan is 5, and Mary is three years older than Juan, Mary must be 8. Notice that since we're using real numbers, it's easy to make sure we're doing the right math: Is 8-year-old Mary 3 years older than 5-year-old Juan? Definitely.

We know Mary's age now, so we can go back to that first part of the problem that we skipped earlier. If George is twice as old as Mary, then George is 16 years old. Now we can answer the question: "If Juan is j years old, then what, *in terms of j*, will be George's age in 10 years?" Ignore the phrase **in terms of.** Whenever you see that in a plugging in problem, it just means that you plug in. If George is 16 now, how old will he be in 10 years? He'll be 26 years old, so that's **our target number.** Write down 26 and circle it.

Now we can check all of the answer choices.

Replace j with 5 in each of the answer choices; the one that gives you an answer of 26 is the correct answer.

(A) $2(5) - 4 = 6$. Nope.

(B) $2(5) - 14 = -4$. Worse than nope, it's impossible; no one can have a negative age.

(C) $2(5) + 13 = 23$. Nope.

(D) $2(5) + 16 = 26$. Bingo!

(E) $2(5) + 26 = 36$. Nope.

Since (D) is the only answer that gave us our target number of 26, that's our answer. Your scratch paper should look something like this:

11)

~~A~~ 2(5) − 4 = 6
~~B~~ 2(5) − 14 = −4
~~C~~ 2(5) + 13 = 23
(D) 2(5) + 16 = 26
~~E~~ 2(5) + 26 = 36

Juan = j = 5
Mary = 8
George = 16
G in 10 years = (26)

2(5) − 14
10 − 14 = −4

If you had solved this algebraically, you might have added j and 3, then doubled it to $2(j + 3)$, then distributed it to become $2j + 6$, then added 10, and gotten the right answer. But say you forgot to distribute the 2, and $2(j + 3)$ became $2j + 3$, and you added 10 to get $2j + 13$. You could have chosen answer choice (C) and moved merrily along, unaware of your mistake, because ETS anticipated it.

Instead, you solved the problem using only arithmetic that you can plug into your calculator, and you avoided all of the traps. Let's try another.

Question 8 of 20

If a machine produces 3,000 golf balls per hour, how many golf balls can four of these machines make in y minutes?

- ○ $200y$
- ○ $\frac{200}{y}$
- ○ $750y$
- ○ $\frac{7500}{y}$
- ○ $\frac{12000}{y}$

Here's How to Crack It

Woah, look at all those variables in the answer choices! Now that we've **recognized the opportunity to plug in**, go ahead and **set up your scratch paper** by writing A B C D E vertically on the left side. Now **choose an easy number to plug in**. There are 60 minutes in an hour, so you can make your math a little easier by choosing either a factor or a multiple of 60, such as 30. Write down $y = 30$.

In 60 minutes, one machine makes 3,000 golf balls. In that case, in 30 minutes one machine would make 1,500 golf balls. The question asks how many golf balls four machines can make in 30 minutes, which is 1,500 × 4 = 6,000 golf balls. Write down 6,000 golf balls and circle it. We've **found our target number**, so we now have to **check all of our answer choices** and see which one gives us 6,000.

Replace x with 30 in each of the answer choices; the one that gives you an answer of 6,000 is the correct answer.

(A) 200(30) = 6,000. Gotcha!

(B) $\frac{200}{30} = 6.66$. Way too small.

(C) 750(30) = 22,500. Way too big.

(D) $\frac{7500}{30} = 250$. Nope.

(E) $\frac{720,000}{30} = 24,000$. Nope.

The answer is (A).

You might be asking yourself, "If I got a match right away, why did I have to spend that time checking all of the others?" And that's a good question, because your ultimate goal is to find the correct answer and scoot off to the next question as quickly as possible. Sometimes we may choose a number that works with some answer choices, but doesn't always work. For instance, say we had plugged in $y = 60$ instead. In that case, the four machines would have made 12,000 golf balls. Answer (A) still works: 200(60) = 12,000. However, look at answer (E): $\frac{72000}{60} = 12,000$. Had we picked 60, we would have had two answer choices that gave us our target number. If that happens, just pick a different number and check the remaining answers.

What to Plug In

When you plug numbers into a question, it's perfectly fine to choose almost any number you want, as long as it doesn't violate restrictions that the problem stipulates. Usually, the first integer that pops into your head will work just fine, although as you get better at it you'll get a feel for picking numbers that make your math easier.

Question 13 of 20

At Markham Academy, $\frac{2}{3}$ of the m faculty members live in housing on campus grounds. Of those living on campus, $\frac{3}{4}$ own a car. Which of the following represents, in terms of m, the number of faculty members who live in campus housing and DO NOT own a car?

- ○ $\frac{m}{4}$
- ○ $\frac{m}{6}$
- ○ $\frac{m}{12}$
- ○ $\frac{2m}{3}$
- ○ $\frac{3m}{4}$

Here's How to Crack It

Since we've got variables in the answer choices, **recognize the opportunity to plug in** and **set up your scratch paper.** Now let's **choose an easy number for the variable.** This question has a lot of fractions in it, and fractions mean division. The fastest way to come up with a good number is to multiply the denominators: Since we know we're going to have to divide by 3 and 4, let's try $m = 12$.

Now we'll work through the problem in bite-sized pieces. Since $\frac{2}{3}$ of the 12 faculty members live on campus, $\frac{2}{3}(12) = \frac{2}{3} \times \frac{12}{1} = \frac{2}{1} \times \frac{4}{1} = 8$ people live on campus. Write down on your scratch paper that 8 people live on campus. The problem then states that of those 8 people living on campus, $\frac{3}{4}$ own a car. $\frac{3}{4} \times \frac{8}{1} = \frac{3}{1} \times \frac{2}{1} = 6$ people on campus own a car. The question asks how many people live on campus DO NOT own a car, which means that out of the 8 people on campus, 2 of them don't own a car. Now that we've **found our target number**, write down 2 on your scratch paper and circle it.

Finally, we have to **check each answer choice**. Replace *m* with 12 in each answer choice to find which one gives us our target number of 2.

(A) $\frac{12}{4} = 3$. Next.

(B) $\frac{12}{6} = 2$. Yes!

(C) $\frac{12}{12} = 1$. Nope.

(D) $\frac{2(12)}{3} = \frac{24}{3} = 8$. Nope.

(E) $\frac{3(12)}{4} = \frac{36}{4} = 9$. Nope.

The answer is (B).

It Gets Easier With Practice

The more problems you work on, the better you'll get at choosing the best numbers. If a problem involves percents, for example, you'll probably plug in a multiple of 100. Questions based on inches and feet might work best with a multiple of 12. If you're dealing with units of time, you might think in multiples of 60. Numbers like these often suggest themselves when you use common sense and ask yourself, "What numbers will make this easier?" And, for that matter, "What numbers will make this harder?"

What *Not* to Plug In

Knowing the right number to choose for a plug-in problem is useful, but it's more important to know what numbers to avoid. These are the numbers that can have a strange effect on the algebra and skew your results.

The following chart shows the numbers that cause trouble when you plug them into equations. These numbers aren't forbidden, but it's best to avoid using them for most problems.

What	Why It's Trouble
0	Additive identity (anything plus 0 equals itself) Anything times 0 equals 0
1	Multiplicative identity (anything times 1 equals itself)
Any numbers that appear in the question	Lots of opportunity for duplicate answers; if $x = 2$, then answer choices "$2x$" and "x^2" both yield a target answer of 4

No Variables? No Problem!

Believe it or not, you can plug into a problem that doesn't appear to have any variables at all. In these cases, there might not be any *x*'s or *y*'s, but there will still be an unknown quantity that, if you knew it, would make the problem easier to solve. Questions like these usually have answer choices that are fractions or percentages.

Question 18 of 20

During the first month after Johanna purchased stock in Amalgamedia, Inc., the value of her shares fell by 20 percent. During the following month, however, Amalgamedia announced record profits and the stock rose 40 percent. By what percent did shares of Amalgamedia change over those two months?

- ○ 4%
- ○ 12%
- ○ 20%
- ○ 50%
- ○ 68%

Here's How to Crack It

There are no variables here, but the problem would make a little more sense if you knew the price of a share of Amalgamedia when Johanna bought into it. So make one up. The added bonus of having no variables is that there's no plugging into the answer choices at the very end, so you can skip Step Three.

Step One: Choose a number for each variable in the problem.

Because percents are involved, let's say the stock cost $100.

Step Two: Come up with a "target answer."

If the stock dropped by 20% during the first month, then price dropped by $\frac{20}{100} \times 100$, or $20, and the price went from $100 to $80. After a 40% increase, the stock moved up by $\frac{40}{100} \times 80$, or $32, to $112. This represents a 12% increase from the original $100, so the answer is (B).

Plugging In Quick Quiz

Question 1 of 4

Joshua is three times as old as Kali, and Kali's age is three years more than double Louella's age. If Joshua is j years old, which of the following represents Louella's age?

○ $\frac{j-3}{6}$

○ $\frac{j+3}{3}$

○ $\frac{j-6}{9}$

○ $\frac{j+9}{3}$

○ $\frac{j-9}{6}$

Question 2 of 4

Ana has a collection of books that are either fiction or non-fiction. Sixty percent of Ana's books are fiction, and 30 percent of the non-fiction books are about politics. What fraction of the books are non-fiction books that are NOT about politics?

○ $\frac{3}{25}$

○ $\frac{7}{25}$

○ $\frac{3}{10}$

○ $\frac{2}{5}$

○ $\frac{3}{5}$

Question 3 of 4

A community park has two rectangular playgrounds—a little one for children younger than age 5, and a larger one for children aged 5 and older. The larger playground is twice as long and five times as wide as the smaller playground. If the area of the small playground is M, then the area of the larger playground is how much larger than the area of the smaller playground?

○ $2M$

○ $5M$

○ $9M$

○ $10M$

○ $11M$

Question 4 of 4

An Internet café offers Internet access at the rates of \$3 per half-hour for the first two hours and \$2 for every half hour after that. If Charlene worked at the café for $x + 5$ hours, how much money, in terms of x, did she have to pay?

- ○ $2x + 18$
- ○ $2x + 15$
- ○ $2x + 12$
- ○ $4x + 24$
- ○ $4x + 48$

Explanations for Plugging In Quick Quiz

1. Let's start with Louella and set $l = 5$. If Kali is three years more than double Louella's age, then $k = 3 + (2 \times 5)$, or 13. Joshua is three times as old as Kali, so $j = 3 \times 13$, or 39. Louella's age (5) is the target, and if you plug 39 in for j among all the answer choices you'll find that (E) is the correct answer: $\frac{39-9}{6} = 5$.
2. There is no variable in the problem, but knowing the number of books Ana has would be a great help. Because we're dealing with percents, let's say she has 100 books. If 60 percent are fiction, then the other 40 are non-fiction. Because 30 percent of those 40, or 12, are about politics, the other 28 are neither fiction nor about politics, and $\frac{28}{100}$ reduces to $\frac{7}{25}$. The answer is (B).
3. If the length and width of the small playground are 8 feet and 10 feet, respectively, then the area of that playground is 8×10, or 80 square feet. Therefore, $M = 80$. The larger playground is twice as long ($2 \times 8 = 16$) and five times as wide ($5 \times 10 = 50$), so its area is 16×50, or 800 square feet. The difference in these areas is $800 - 80$, or 720 square feet (target answer). When you plug in $M = 80$ to each answer choice, you'll see that the answer is (C). If you leapt right toward (D), you misread the question. Slow down and make sure you understand each question completely!
4. We don't want to choose a value of x that we've already seen in the question or answer choices, so let $x = 6$. That means Charlene worked for 11 hours. She paid \$12 for the first two hours (4 half-hours at \$3 each), and the remaining nine hours (18 half-hours at \$2 each) cost \$36. The total (and target answer) is $12 + 36$, or \$48. After you plug in $x = 6$ to the answer choices, you'll find the answer is (D).

"Must Be" Problems

Every so often you'll see a question that contains variables and the phrase "must be." Basically, you can eliminate an answer choice as soon as you find one situation when it doesn't work.

Because Must Be problems are looking for the answer choice that always works, we may need to try a lot of different numbers until we're down to only one answer. The trick here is to use the numbers that most people don't think to use. We'll call those numbers ZONEF.

Z – Zero

O – One

N – Negative numbers

E – Extremely large or small numbers

F – Fractions or decimals

Trigger: Variables in the answer; problem says "must be."

Response: Plug in a simple number, and then use ZONEF numbers.

We won't have to try every single ZONEF number for Must Be problems, but we may have to try several. First, however, we'll try an easy number. It may be our usual easy numbers: 2, 3, 5, 10, or 100, but it could be any number that seems easy for the question. Don't think too hard about finding the perfect number at this point. We're just going to use the easy number to eliminate some or most of the answers. Once we've eliminated some answers, we'll plug in using a ZONEF number and see which other answers we can eliminate.

1. **Recognize the Opportunity to Plug In.** If you have variables in the answer choices and the problem says "must be," plug in.
2. **Set Up Your Scratch Paper.** Write down letters for each answer choice and any important information from the problem.
3. **Plug In an Easy Number for the Variable.** Choose an easy number, such as 2, 3, 5, 10, or 100, for one of the variables. If the problem gives you certain limitations for the value of the variable, such as "odd integer larger than 30," be sure to choose a number that follows those limitations. If there are multiple variables in a problem, see if you can solve for the other variables once you've plugged in for the first one. If you can't, plug in for those variables as well.
4. **Check All of the Answer Choices.** You're not looking for the right answer, but simply looking for any answers that you can eliminate.
5. **Try a Weird ZONEF Number.** Try Zero, One, Negative numbers, Extremely large or small numbers, or Fractions. Check all the remaining answers. If necessary, try another ZONEF number.

Question 10 of 20

If $|c| < 4$ and $d = 3c - 2$, which of the following must be true?

- ○ $-2 \leq d < 10$
- ○ $d \neq -2$
- ○ $d < 0$
- ○ $c < d$
- ○ $-14 < d$

Here's How to Crack It

First off, **recognize the opportunity to plug in.** We've got variables in all the answer choices, and the question says "must be." **Set up your scratch paper** by writing down A B C D E. Now let's **try an easy number.** If $c = 2$, then $d = 3(2) - 2 = 6 - 2 = 4$. Now let's **check each answer.** (A) works, because $-2 \leq 4 < 10$, and (B) works, because $4 \neq -2$. Cross off (C), because 4 is not less than 0. (D) works, because $2 < 4$, and (E) works, because $-14 < 2$. We've still got 4 answer choices left, so let's **try a ZONEF number.** Let's start with $c = 0$. If $c = 0$, then $d = 3(0) - 2 = -2$. Leave (A) because d is equal to -2. Cross off (B) because d does, in fact, equal -2. We've already eliminated (C), so we don't have to check it again. We can eliminate (D), because $0 < -2$. We can leave (E), because $-14 < -2$. Now let's try another ZONEF number. Since there's an absolute value in the problem, let's try a negative number for c. If $c = -3$, then $d = 3(-3) - 2 = -9 - 2 = -11$. We've only got (A) and (E) left, so let's try those. Cross off (A) because -3 is smaller than -2. The only answer left is (E), which works: -14 is less than -11. The answer is (E).

Here's what your scratch paper should look like:

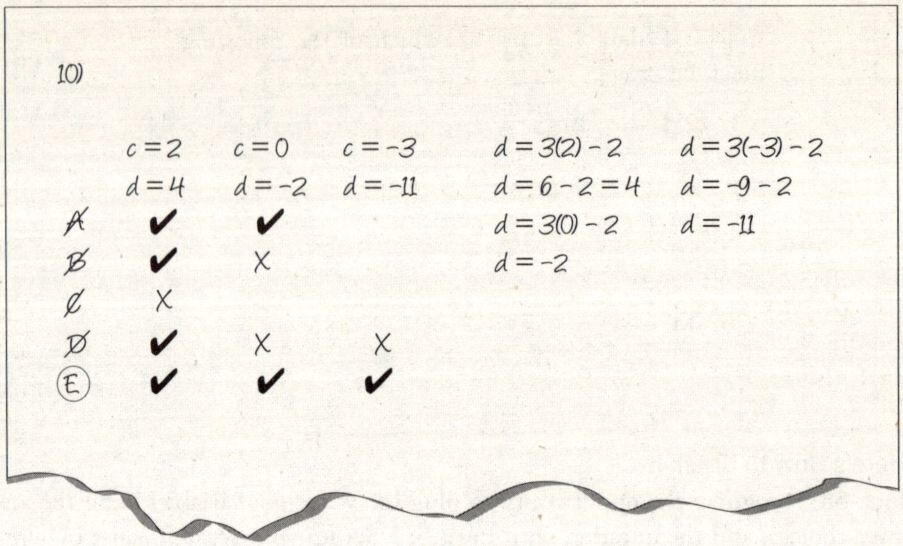

ZONEF numbers are weird numbers. These are the numbers people normally don't think of when working on GRE math problems. However, there may be other numbers other than the ZONEF numbers that work better for certain problems. For instance, a problem about even and odd numbers may require plugging in an even number and then an odd number. We may need to try prime numbers or perfect squares on a problem about factoring. ZONEF will work most of the time, so stick with it unless you definitely see a different type of number that is necessary for that problem.

"Must Be" Quick Quiz

Question 1 of 4

If $xy \neq 0$ and y is even, which of the following must also be even?

○ $\frac{2x}{y}$

○ $\frac{y}{2} + x$

○ $3y - 2x$

○ $\frac{y}{x} + \frac{x}{y}$

○ $2y - 3x$

Question 2 of 4

If b and c are negative integers, which of the following must also be negative?

- ○ $b^3 - c^3$
- ○ $bc^2 - c$
- ○ $bc(b - c)$
- ○ $\frac{b}{c} + \frac{c}{b}$
- ○ $b^2c + bc^2$

Question 3 of 4

If $m > 0 > n$ and m is odd, all of the following could be odd and positive EXCEPT

- ○ $(mn)^2$
- ○ $n^3 - m^3$
- ○ $(m - n)^2$
- ○ $m^2 + n^2$
- ○ $m(m + n)$

Question 4 of 4

If a positive integer a is multiplied by an integer b and the result is less than a, then it must be true that b is

- ○ equal to 1
- ○ greater than 0
- ○ greater than a
- ○ greater than 1
- ○ less than 1

Explanations for "Must Be" Quick Quiz

1. If $xy \neq 0$ and y is even, you can set $x = 2$ and $y = 4$ and consider the results.

   $\frac{2(2)}{4} = 1$, which is odd. Eliminate it.

   $\frac{4}{2} + 2 = 4$, which is even. Keep it.

   $3(4) - 2(2) = 8$, which is even. Keep it.

   $\frac{4}{2} + \frac{2}{4} = 2\frac{1}{2}$, which you can get rid of because it isn't an integer.

   $2(4) - 3(2) = 2$, so you can keep this as well.

   Now, the only leeway for change that you have is to make x odd. Let's make $x = 3$, leave $y = 4$, and check the remaining three.

   $\frac{4}{2} + 3 = 5$, which is now odd. Get rid of it.

   $3(4) - 2(3) = 2$, which is still even. Keep it.

   $2(4) - 3(3) = -1$, which is odd.

   You've eliminated all the others, so the answer is (C).

2. Plug in two negative numbers, and see what happens. Let $b = -2$ and $c = -3$.

   $(-2)^3 - (-3)^3 = -8 + 27$, which equals 19. Dump it.

   $(-2)(3)^2 - (-3) = -18 + 3 = -15$. Keep it.

   $(-2)(-3)[-2 - (-3)] = 6$. Dump it.

   $\frac{-2}{-3} + \frac{-3}{-2} = \frac{13}{6}$. Dump it.

   $(-2)^2(-3) + (-2)(-3)^2 = -12 + -18 = -30$. Keep it.

   We have two answer choices left, and we have to find a way to make one of them positive. If you let b and c both equal -1, you get what you're looking for: $(-1)(-1)^2 - (-1) = 0$, which isn't negative, while $(-1)^2(-1) + (-1)(-1)^2 = -2$. The answer is (E).

3. Because m is odd and positive, let $m = 3$. Now, let's plug in $n = -2$ and see what transpires.
$(3 \times -2)^2 = 36$. Positive but not odd, so get rid it.
$(-2)^3 - 3^3 = -35$. Odd but not positive, so keep it.
$[3 - (-2)]^2 = 25$ Odd and positive, so get rid of it.
$3^2 + (-2)^2 = 13$. Also gone.
$3(3 + -2) = 3$. Ditto.
You have two answers left, and you want to make answer choice (A) odd or answer choice (B) positive. If you plug in an odd number for n in (A), you can do the former: $(3 \times -3)^2 = 81$, which is odd and positive. There's no way (B) will ever be positive, so the answer is (B).
4. Think of this one from a slightly different perspective by choosing a value of a and the value after a is multiplied by b; this product ab must be less than a. But, if you multiply a positive integer by another positive integer, the product will always be the same or bigger (Try plugging in 2 for a!). So you can eliminate answer choices (A), (B), (C), and (D). But no one said that b was positive. Therefore, if $a = 2$ and $b = 3$, it's possible for ab to equal -6, which is less than 2. The answer is (E).

Plugging In on Quant Comps

Plugging In for Quant Comp questions is similar to Plugging In on Must Be questions. We've got four possible answers: (A), (B), (C), and (D). As we try different numbers, we'll eliminate answers. We'll want to pick some easy numbers first, and then try our weird ZONEF numbers until we've either eliminated all the answers but (D) or keep getting (A), (B), or (C).

Let's review the answer choices for Quant Comp questions first.

(A) means that Quantity A is *always* larger than Quantity B. B is never larger than A, and they are never the same.
(B) means that Quantity B is *always* larger than Quantity A. A is never larger than B, and they are never the same.
(C) means the two quantities are *always* equal. Quantity A is never larger than B, and Quantity B is never larger than A.
(D) means we're not sure. Sometimes Quantity A is larger, sometimes B is larger, or sometimes they're the same.

As we try each number, we'll check our quantities. Is Quantity A larger right now? Then eliminate answers (B) and (C), because Quantity B is not always larger than A, and the two quantities are not always the same. Is Quantity B larger right now? Then eliminate (A) and (C), because Quantity A isn't larger, and they're not the same. Are the two quantities the same? Then eliminate (A) and (B), because Quantity A isn't always larger, and Quantity B isn't always larger.

So once we know one possible value for Quantity A and B, we can cross off two answers. Already down to a 50-50 shot, just by plugging in one number! Then we'll try to plug in more numbers, and see if we can eliminate whichever of (A), (B), or (C) is left.

Trigger: Quant Comp with variables.

Response: Set-up your scratch paper and plug in using ZONEF.

Notice that we'll never eliminate (D) when we plug in. Answer choice (D) is our last ditch effort. We don't want to pick (D), so first we're going to try as hard as possible to make sure (A), (B), or (C) work. If we eliminate all those, however, then we're stuck with (D).

The Steps in Detail

- **Use your scratch paper:** Write down the question number, A B C D vertically on the left side of the paper, and any information given in the problem. Remember that we're going to mostly use POE with Quant Comp plug in questions, so writing down the answer choices is incredibly important.
- **Recognize that it's a Plug In:** If it's a Quant Comp question with variables in both the answers, it's definitely a plug in question. You should immediately know how you're going to solve that question: by Plugging In! If one quantity has a variable in it, and the other quantity is a number, then it may be a plug in. Check to see if there's any quantity referenced in the problem that you can't actually solve for. If so, it's a plug in question.
- **Try an easy number:** Choose a nice, easy number to plug in for your variable. Don't worry about being clever here, and don't spend time thinking of the perfect number. Just try out whatever number you can think of that is allowed by the question. 2, 3, 5, 10, or 100 are all easy numbers to try out. Work through the problem with your easy number. If you have multiple variables in the question, see if you can solve for the other variables once you've plugged in one of them: If not, you'll have to plug in for that variable as well.
- **Cross off two answers:** Once you've solved the question with your first number, compare Quantity A and Quantity B. If A is bigger, cross off (B) and (C). If B is bigger, cross off (A) and (C). If the quantities are the same, cross off (A) and (B).
- **Try a ZONEF number:** Now we're down to two answer choices, and we want to make sure that whichever answer we have, (A), (B), or (C), always works, no matter what we throw at the problem. Once again, our ZONEF numbers are great here, although those aren't the only possibilities. If Quantity A is larger, try any number that you think will make Quantity B larger (and vice versa).
- **If you can't eliminate (A), (B), or (C), choose that.** You may have to

try several different numbers until you can prove to yourself that the answer is always (A), (B), or (C). If none of the ZONEF numbers look like they'll change the quantities much, and there aren't any other numbers you can try, then go ahead and pick which answer, (A), (B), or (C), you keep getting.

- **If the only answer left is (D), pick (D).** Crossed off (A), (B), and (C)? Then you're stuck with poor, lonely (D). Looks like you'll have to pick it.

Question 6 of 20

Quantity A	**Quantity B**
$2x + 1$	$3x + 1$

○ Quantity A is greater.

○ Quantity B is greater.

○ The two quantities are equal.

○ The relationship cannot be determined from the information given.

At first glance, you might think instinctively that Quantity B must always be greater, because 3 is bigger than 2. The GRE is engineered to take advantage of these instincts, however. Follow our steps and you'll see what we mean.

Start by **setting up your scratch paper.** Write down A B C D, and copy down Quantity A and Quantity B. Any scratch work we have to do, such as working out expressions for each quantity, will be done on the right side of the paper.

Choose an easy number. Let's try $x = 3$. Write down $x = 3$ between the two columns. Now let's find the values of each quantity. Quantity A is $2(3) + 1 = 6 + 1 = 7$. Quantity B is $3(3) + 1 = 9 + 1 = 10$. Right now, B is larger than A, so we can eliminate (A) and (C).

Now let's see if we can pick a number that makes it so that Quantity B is no longer larger. **Check your ZONEF numbers:** How about zero? Quantity A is $2(0) + 1 = 0 + 1 = 1$. Quantity B is $3(0) + 1 = 0 + 1 = 1$. Now Quantity A equals Quantity B, which means B isn't always larger. Cross off (B). Looks like we're stuck with (D), which is the answer.

Your scratch paper should look something like this:

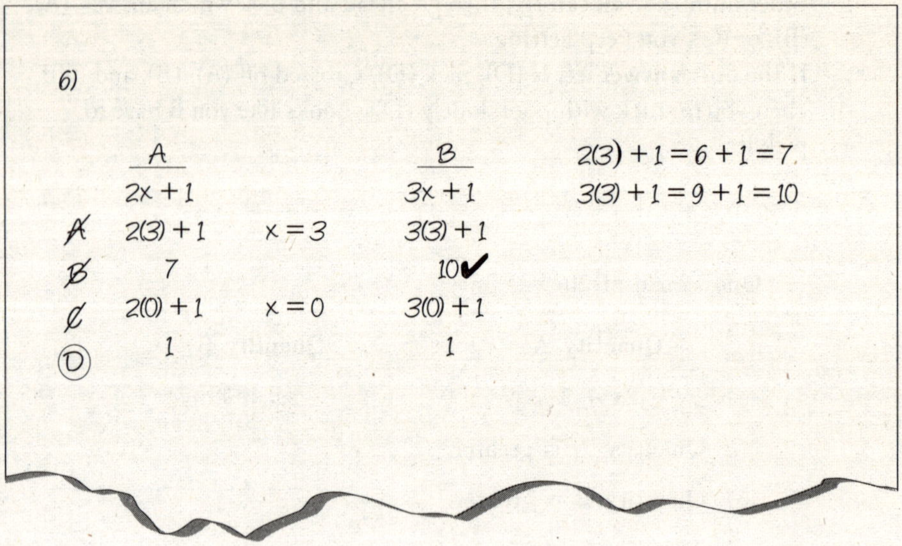

See how this works? Rather than guess, you can use a simple, methodical approach to eliminating the three wrong choices and choosing the one that remains. Let's try another:

Question 5 of 20

Quantity A	**Quantity B**
$1000x + 10$	x^2

- ○ Quantity A is greater.
- ○ Quantity B is greater.
- ○ The two quantities are equal.
- ○ The relationship cannot be determined from the information given.

Here's How to Crack It

Here is another situation in which instinct might lead you down the wrong road. Look how big Quantity A looks! How could Quantity B possibly measure up?

As always, get your hand moving by **setting up your scratch paper.** Write down A B C D, and copy down Quantity A and Quantity B. Now we're ready to **choose an easy number.** Let's try $x = 2$. Quantity A is $1{,}000x + 10$, so it's $2{,}000 + 10 = 2{,}010$. Quantity B is $2^2 = 4$. Since Quantity A is definitely larger than Quantity B, cross off answers (B) and (C).

We're left with (A) and (D), so now let's **try ZONEF.** We don't have to try every single number, but let's look to see if any of them look like they'll make Quantity B larger than A. We could try zero again, but this time it just confirms what we already know: If $x = 0$, Quantity A would equal $0 + 10 = 10$ and Quantity B would be 0, so Quantity A is still larger.

We've tried two sets of numbers now, but we're still not done. Keep looking over ZONEF to see if any other numbers look promising. How about negative numbers? Let's try $x = -5$. Quantity A is $1{,}000(-5) + 10 = -5{,}000 + 10 = -4{,}990$. Quantity B is $(-5)^2 = 25$. Since A is negative and B is positive, Quantity B is now larger, and we can eliminate (A). Our answer is (D).

Here's what your scratch paper should look like for this problem:

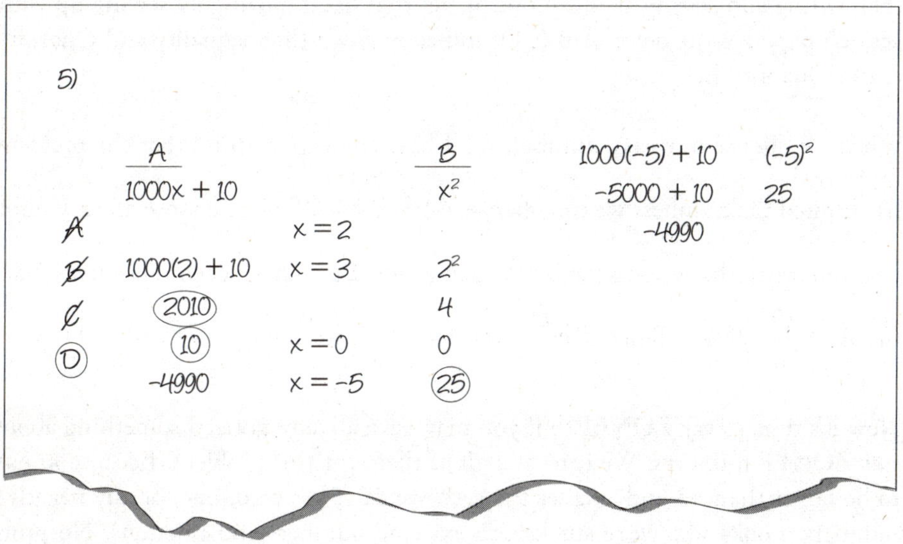

Sometimes, we'll have limitations on what we can plug in for certain variables. Whenever we plug in, we'll have to check to make sure that the numbers we plug in are allowed by the problem.

Question 2 of 20

$$42 < m < 49$$

Quantity A	Quantity B
$\frac{m}{56}$	0.75

○ Quantity A is greater.

○ Quantity B is greater.

○ The two quantities are equal.

○ The relationship cannot be determined from the information given.

Here's How to Crack It

We've got a lot more pieces of information to deal with on this problem, so rather than sitting and staring at this problem, get that hand moving by **setting up your scratch paper**. Write down A B C D, and copy down the inequality and Quantity A and Quantity B.

When we **choose our easy number**, we'll have to keep in mind that the problem has limited the numbers we can choose: $42 < m < 49$. So let's choose an easy number somewhere between 42 and 49, such as 45. Let's check each quantity. Quantity A is $\frac{45}{56} \approx 0.803$. Since that's larger than 0.75, we can cross off (B) and (C).

Now it's time to **try ZONEF**, but you may have already noticed something about our ZONEF numbers: We can't use all of them for this problem. Because *m* has to be larger than 42 and smaller than 49, we can't use zero, one, or any negative numbers. Looks like we're stuck with extreme numbers and fractions. No problem. We'll use a number either as extremely large or extremely small as is possible within our limited range for *m*, and we'll use some non-integers.

Before we choose another number, however, let's look at what we want to get. Right now Quantity A is bigger than B, and we want to see if we can make it smaller. To make Quantity A smaller, we'll want to choose as small a number as possible for m. Since we're limited by $42 < m < 49$, let's choose the smallest number we can for m, such as $m = 42.001$. We could get smaller, but let's start there for now. $\frac{42.001}{56} \approx 0.7500179$. Close, but still a little bigger than B, so we can't eliminate any answers. Let's go even smaller then, and try $m = 42.00001$, which is about as much as we can enter into the on-screen calculator. $\frac{42.00001}{56} \approx 0.7500002$. A is still larger than B and nothing we pick seems to change that. We're down to (A) and (D), and since we couldn't eliminate (A), that's our answer.

As usual, here's an example of what your scratch paper should look like. Notice that for each fraction entered into the calculator, we wrote down the fraction and the result of the calculation on the right side of our scratch paper.

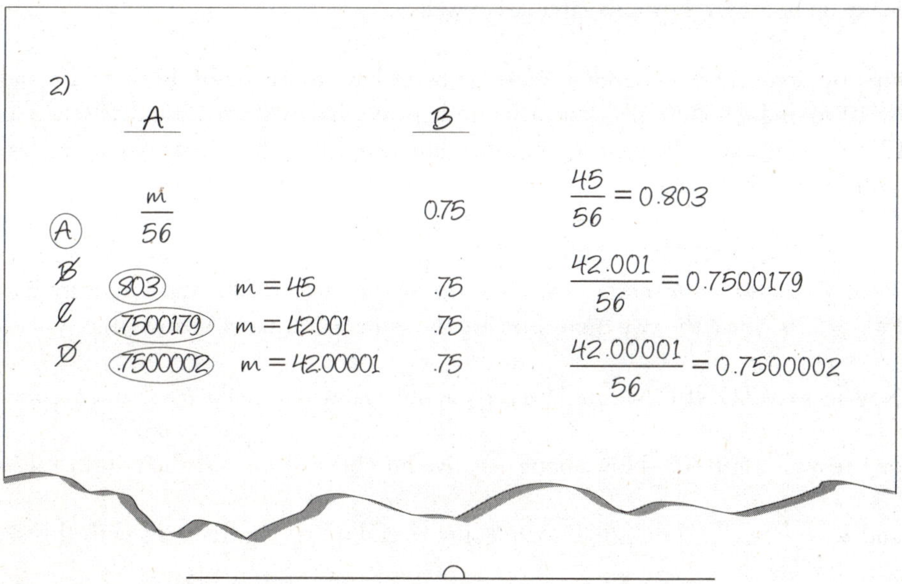

If we have multiple variables, then we'll only plug in for one variable at a time. If we can't solve for any of the other variables, then we'll plug in a different number for the next variable, and see if we can solve for any of the other unknowns.

Question 4 of 20

$$4a = 12b$$
$$2b = 10c$$

Quantity A	Quantity B
a	$15c$

- ○ Quantity A is greater.
- ○ Quantity B is greater.
- ○ The two quantities are equal.
- ○ The relationship cannot be determined from the information given.

Here's How to Crack It

Variables in both answer choices mean you probably **noticed that we can plug in.** Pick up your pen and start **setting up your scratch paper.** Write down A B C D and copy down the two equations, Quantity A, and Quantity B. We've got three variables in this problem, which means we'll still **choose an easy number,** but we'll then have to see if we can solve for the other variables. If we can't, then we'll make up numbers for those variables as well.

First off, look at the equations. We're going to have to divide by 12 to find b and by 10 to find c, so let's pick an easy number divisible by both 12 and 10, such as 120. We could actually pick any number, but 120 will keep us from having to deal with too many fractions.

If $a = 120$, then $4(120) = 12b$, and $480 = 12b$, and $b = 40$. Now use b to find c, so $2(40) = 10c$, $80 = 10c$, and $c = 8$. Quantity A is therefore 120, and Quantity B is $15(8) = 120$. Since the two quantities are the same, eliminate (A) and (B).

Now let's **try ZONEF.** We could try $a = 0$, but that would make $b = 0$ and $c = 0$, so our answer is still (C). How about negative numbers? If $a = -2$, then $4(-2) = 12b$, and $b = \frac{-8}{12} = -\frac{2}{3}$. Looks like we're going to end up trying fractions with this set of numbers. Since $2\left(-\frac{2}{3}\right) = 10c$, $-\frac{4}{3} = 10c$ and $c = -\frac{4}{30} = -\frac{2}{15}$. Quantity A is -2, and Quantity B is $15\left(-\frac{2}{15}\right) = -2$, our answer is still (C). Since we can't find any numbers to eliminate (C), that's our answer.

Scratch paper:

4)	<u>A</u>	$4a = 12b$	<u>B</u>	$4(120) = 12b$	$4(-2) = 12b$
		$2b = 10c$		$480 = 12b$	$-8 = 12b$
	a		$15c$	$40 = b$	$-\frac{8}{12} = b$
		$a = 120$			
~~A~~	120	$b = 40$	$15(8)$	$2(40) = 10c$	$-\frac{2}{3} = b$
~~B~~		$c = 8$	120	$80 = 10c$	
(C)				$c = 8$	$2\left(-\frac{2}{3}\right) = 10c$
D		$a = 0$			
	0	$b = 0$	0		
		$c = 0$			$-\frac{4}{3} = 10c$
		$a = -2$			$-\frac{4}{30} = c$
	-2	$b = -\frac{2}{3}$	$15\left(-\frac{2}{15}\right)$		$-\frac{2}{15} = c$
		$c = -\frac{2}{15}$	-2		

As we plug in for multiple variables, plug in different numbers for each variable on your first attempt.

Question 5 of 20

Quantity A	**Quantity B**
ab	$a + b$

- ○ Quantity A is greater.
- ○ Quantity B is greater.
- ○ The two quantities are equal.
- ○ The relationship cannot be determined from the information given.

Here's How to Crack It

Pick up your pen and set up your scratch paper: We've got variables in our answers, which means this is a plug in problem. Once you've done that, we need to figure out two different easy numbers: Let's say $a = 2$. Does that tell us what b is? Nope, we don't have any way of figuring out b, so let's plug in for b as well, and say $b = 3$. Quantity A is $2 \times 3 = 6$. Quantity B is $2 + 3 = 5$. Quantity A is bigger, so cross off (B) and (C). Do any of our ZONEF numbers seem promising? We could try negative numbers, but if we try $a = -2$ and $b = -3$, then Quantity A is $(-2)(-3) = 6$ and B is $(-2) + (-3) = -5$, and A is still larger. But who says we have to make both numbers negative? Let's try $a = -4$ and $b = 5$. Now Quantity A is $(-4)(5) = -20$, and Quantity B is $(-4) + (5) = 1$, and Quantity B is larger, so we can cross off answer (A). Our only answer left is (D).

Be Vigilant

When you're focusing on the two quantities, it's easy to miss the text immediately above. You wouldn't think it would happen often, but it does. So keep a watchful eye and make sure that everything you Plug In to the Quant Comp problem is nice and legal.

Quant Comp Plugging In Quick Quiz

Question 1 of 6

The integer k is negative.

Quantity A	**Quantity B**
$5 - 2k$	$2 - 5k$

- ○ Quantity A is greater.
- ○ Quantity B is greater.
- ○ The two quantities are equal.
- ○ The relationship cannot be determined from the information given.

Question 2 of 6

Quantity A	Quantity B
$(b - 15)(b + 15)$	$225 - b^2$

- ○ Quantity A is greater.
- ○ Quantity B is greater.
- ○ The two quantities are equal.
- ○ The relationship cannot be determined from the information given.

Question 3 of 6

The integer h is positive.

Quantity A	Quantity B
$h^2 - 16$	$h^2 + h - 12$

- ○ Quantity A is greater.
- ○ Quantity B is greater.
- ○ The two quantities are equal.
- ○ The relationship cannot be determined from the information given.

Question 4 of 6

w, x, y, and z are consecutive, positive, even integers, and $w < x < y < z$

Quantity A	Quantity B
$2(w + x) + 2$	$y + z - 2$

- ○ Quantity A is greater.
- ○ Quantity B is greater.
- ○ The two quantities are equal.
- ○ The relationship cannot be determined from the information given.

Question 5 of 6

Quantity A	Quantity B
The cost of m books, each of which costs $n - 3$ dollars	The cost of n books, each of which costs $m + 2$ dollars

- ○ Quantity A is greater.
- ○ Quantity B is greater.
- ○ The two quantities are equal.
- ○ The relationship cannot be determined from the information given.

Question 6 of 6

Albert is a inches tall, Beatrice is b inches tall, and Charlene is c inches tall, and c is the average of a and b.

Quantity A	Quantity B
The combined height, in feet, of all three people	$\frac{c}{4}$

- ○ Quantity A is greater.
- ○ Quantity B is greater.
- ○ The two quantities are equal.
- ○ The relationship cannot be determined from the information given.

Explanations for Plugging In Quick Quiz

1. We've got variables in our answer choices, which means this is a plug in question. Set up your scratch paper, and let's try some numbers. Let's start with an easy negative number: $k = -2$. In that case, Quantity A is $5 - 2(-2) = 5 + 4 = 9$. Quantity B is $2 - 5(-2) = 2 + 10 = 12$, so eliminate (A) and (C). Let's try a ZONEF number. We're stuck with negative numbers, so how about $k = -1$? Quantity (A) is $5 - 2(-1) = 5 + 2 = 7$, and Quantity B is $2 - 5(-1) = 2 + 5 = 7$, so both quantities are equal. Eliminate (B), and the answer is (D).
2. Set up your scratch paper with A B C D, Quantity A, and Quantity B. Start with something easy: $b = 5$. Quantity A is $(5 - 15)(15 + 5) = (-10)(20) = -200$. Quantity B is $225 - (5)^2 = 225 - 25 = 200$. Quantity B is larger, so eliminate (A) and (C). Since we want Quantity B to be smaller, we want b

to be bigger, so let's try an extremely large number, such as 100. Quantity A is $(100 - 15)(100 + 15) = (85)(115) = 9{,}775$. Quantity B is $225 - 100^2 = 225 - 10{,}000 = -9{,}775$. Now Quantity A is larger, so we can eliminate (B), and the answer is (D).

3. Variables in both the answers? Pick up your pen and set up your scratch paper. Now that your hand is moving, let's pick an easy number for h. If $h = 2$, then Quantity A is $(2)^2 - 16 = 4 - 16 = -12$. Quantity B is $2^2 + 2 - 12 = 4 + 2 - 12 = 6 - 12 = -6$. Quantity B is therefore larger (remember that negative numbers closer to zero are larger than negative numbers farther from zero), so eliminate (A) and (C). Now let's try a ZONEF number. Since h has to be positive, we can't pick negative numbers or zero, so let's try an extremely large number. If $h = 100$, then Quantity A is $100^2 - 16 = 10{,}000 - 16 = 9{,}984$. Quantity B is $100^2 + 100 - 12 = 10{,}000 + 100 - 12 = 10{,}100 - 12 = 10{,}088$. Quantity B is larger, and nothing seems to change that, so (B) is our answer.
4. Plug in values for the four variables according to the directions. If we let $w = 2$, $x = 4$, $y = 6$, and $z = 8$, then Quantity A equals $2(2 + 4) + 2$, or 14, and Quantity B equals $6 + 8 - 2$, or 12. Quantity A is greater, so eliminate answer choices (B) and (C). Now try greater numbers, like 20, 22, 24, and 26. Quantity A becomes $2(20 + 22) + 2$, or 86, while Quantity B is $24 + 26 - 2$, or 48. Quantity A has become greater, so there is no definitive solution. The answer is (D).
5. You might think the answer to this question is (D), because you don't know the values of m and n. Plugging in numbers, however, reveals a different picture. Say $m = 4$ and $n = 8$; Quantity A becomes the price of 4 books at \$5 each, or \$20, while Quantity B becomes the price of 8 books at \$6, or \$48. Because B is greater, you can eliminate (A) and (C). The only numbers that can make Quantity A larger are negative numbers, and you can't plug in negatives or zero because we're dealing with an actual number of items, and you can't have a negative number of books or a negative price. The answer is (B).
6. Don't just plug in three values for a, b, and c, because the numbers are related. You can pick whatever you want for a and b (such as $a = 36$ and $b = 48$), but c has to be the average of these values. Therefore, $c = \frac{36+48}{2} + 42$. The combined height of all three people is $36 + 48 + 42$, or 126 inches. To convert to feet, divide by 12; $126 \div 12 = 10.5$. Because Quantity B is $\frac{42}{4}$, or 10.5, the answer is (C).

PLUGGING IN THE ANSWER CHOICES

It's great when we can supply our own numbers to the Plug-In party, but sometimes we don't even have to exert that much effort. Often, the GRE is courteous enough to supply five numbers that we can use to plug in. Awfully sporting, don't you think?

Trigger: "How much," "How many," "What is the value," numbers in the answer choices.

Response: Plug in the answers.

This isn't the limit of the GRE's largesse, because as we mentioned before, the answer choices are always listed in numerical order. In order to eliminate the incorrect answer choices most efficiently, you should "Plug In the Answers," (PITA), starting with answer choice (C), because it's in the middle.

The steps for PITA are:

1. **Recognize the opportunity to PITA.** Questions which ask for the value of something (how much, how many, what is the value, et cetera) are PITA questions. If the answer choices are numbers, you can probably PITA.
2. **Write out the answer choices on your scratch paper.** Write the numbers for each answer choice as well as the usual vertical A B C D E.
3. **Label your answer choices.** If the question asks for the number of hats David has, label the column of answer choices as "David's hat." If it's asking for the value of x, label the column of answer choices with an x on top.
4. **Use answer choice (C).** Pretend that the answer to the question is whatever number answer choice (C) is. If you have more or fewer than five answer choices, use the middle answer choice.
5. **Step through the problem.** Use (C) to work through the problem in bite-size pieces. As you come up with values for each step, make a new column next to your answer choices.
6. **Check your answer.** If you end up with numbers that don't match up, then (C) is incorrect. If (C) was too small, cross off any answers smaller than (C), usually (A) and (B). If (C) was too big, cross off any answers bigger than (C), usually (D) and (E). Check one of the remaining answers.

If you're not sure whether your answer was too big or too small, then cross off (C) and pick another answer choice. Once you've tried, say, (B), you'll notice that (B), if it was incorrect, was either closer to the answer you wanted than (C) was, in which case (A) is the answer, or farther away from the answer you wanted, in which case try (D) or (E).

Question 16 of 20

When three consecutive integers are multiplied together, the result is 17,550. What is the largest number in this series?

- ○ 24
- ○ 25
- ○ 26
- ○ 27
- ○ 28

Here's How to Crack It

It seems like there is some algebra we could do to solve this problem, but it may not be obvious what that algebra is. That's fine. In fact, we should avoid using algebra on this problem. It's exactly what ETS knows most people will do, and most people will subsequently make algebraic mistakes.

This question wants to know the largest number, and we've got numbers in the answers. **Recognize the opportunity to PITA** and **write down your answer choices.** Since the question is asking for the largest number, label the column of answer choices "Largest number."

Let's **use answer choice** (C) first. If the largest number in the series is 26, then what are the other numbers? Now it's time to **step through the problem**. The problem states that we have three consecutive integers. If the largest integer is 26, then the other two must be 24 and 25.

To **check our answer**, we'll have to use the last part of the problem: When the numbers are multiplied together, the result is 17,550. Pull up the calculator and multiply together our three numbers. Remember to do one step at a time with the calculator, and write down the result at each stage. $24 \times 25 \times 26 = 600 \times 26 = 15{,}600$. We wanted our result to be 17,550, so the answer we checked was too small. Therefore, we can cross off (C), because it's wrong, and also (A) and (B): If (C) was too small, then (A) and (B) must be too small as well.

We've got (D) and (E) left. Which one should we check? It doesn't matter. If we check (D) and it's wrong, the answer is (E). If we check (D) and it's correct, then we're done. The same is true of (E). Let's try (D). If the largest number is 27, then the other two consecutive integers must be 25 and 26. $25 \times 26 \times 27 = 650 \times 27 = 17{,}550$, which is exactly what we wanted. The answer is (D).

Here's our scratch paper for this problem:

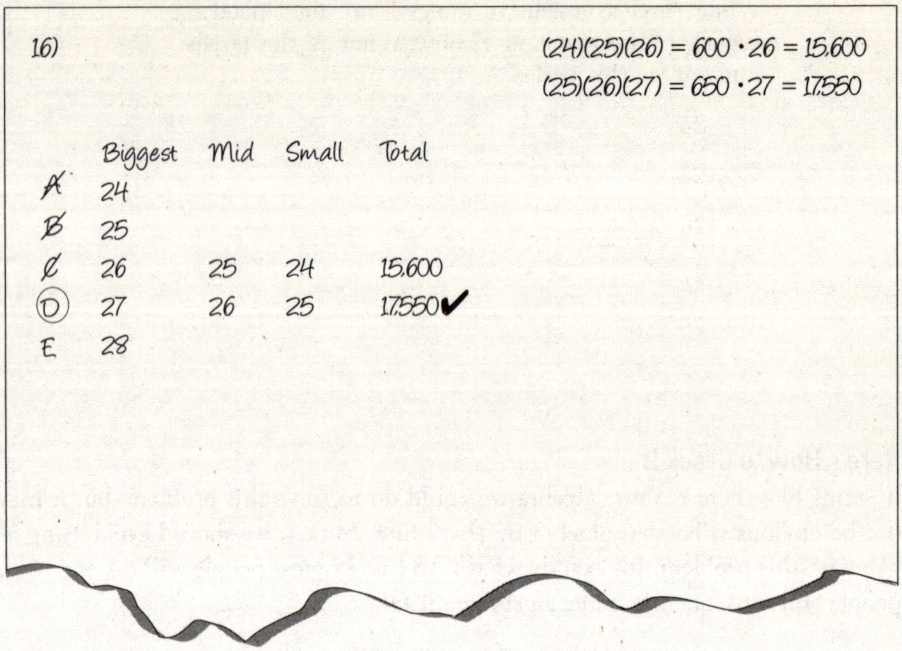

Let's try another one:

Question 12 of 20

Bert and Ernie collect rare, antique waffle irons, and as of last week Bert had twice as many waffle irons as Ernie did. If Bert gave Ernie 6 waffle irons as a gift, and now Bert has 10 more waffle irons than Ernie does, how many waffle irons did Ernie have initially?

- ○ 20
- ○ 22
- ○ 28
- ○ 38
- ○ 44

Here's How to Crack It

Again, start with (C). If Ernie had 28 waffle irons initially, then Bert had twice as many, or 56. Bert then gave Ernie 6 waffle irons and had 50 left; Ernie now has 6 more waffle irons, or 34. The difference between 50 and 34 is 16, which is greater than 10. So 28 is out (answer (C) can be eliminated), and because it's too big, we can eliminate (D) and (E) as well.

If Ernie had 20 waffle irons, then Bert had 40, and after the transaction, Ernie had 26 and Bert had 34. Those numbers are only 8 apart, so answer choice (A) is too small. You've eliminated everything else, so there's nothing left to do. The answer is (B).

PITA Quick Quiz

Question 1 of 4

If $3x^2 + 5x - 2 = 0$, what is a possible value of x ?

- ○ –2
- ○ –1
- ○ 0
- ○ 1
- ○ 2

Question 2 of 4

If $3^x = 9^{(x-4)}$, then $x =$

- ○ 2
- ○ 4
- ○ 5
- ○ 8
- ○ 16

Question 3 of 4

Margarita bought x identical sweaters for \$300. If each sweater had cost \$7.50 less, she could have bought $x + 2$ sweaters for the same amount of money. How many sweaters did she buy?

- ○ 6
- ○ 8
- ○ 10
- ○ 12
- ○ 14

Question 4 of 4

At a certain car dealership, $\frac{2}{3}$ of the cars are four-door sedans and $\frac{2}{5}$ of the rest are SUVs. If the remaining 24 vehicles are pickup trucks, how many vehicles are at the car dealership?

- ○ 60
- ○ 90
- ○ 120
- ○ 180
- ○ 240

Explanations for PITA Quick Quiz

1. Because $3(-2)^2 + 5(-2) - 2 = 0$, the correct answer is (A).
2. You'll need your calculator for this one. Because $3^8 = 6{,}561$ and $9^{(8-4)} = 9^4 = 6{,}561$, the answer is (D).
3. Try (C) first. If Margarita bought 10 identical sweaters for \$300, then they cost \$30 each. Two sweaters more would equal 12 sweaters, each of which would cost $\frac{300}{12}$, or \$25. Because \$30 and \$25 are not \$7.50 apart, you can eliminate (C) and move on. But which way? You want to gap between prices to be wider, so try a smaller number. If she bought 8 sweaters, then they cost \$37.50 each. An extra 2 sweaters would make 10, which we know would cost \$30 each. The difference is \$7.50, and you're work is complete. The answer is (B).

4. If the car dealership has 120 vehicles, then $\frac{2}{3}$ of those 120, or 80, are four-door sedans. Of the remaining 40, $\frac{2}{5}$ are SUVs. Because $\frac{2}{5} \times 40$ is 16, the number of cars that remain to be categorized is 40 – 16, or 24. Bingo! The answer is (C).

PITA also works on the other question types. With All That Apply questions, we'll typically be looking for the biggest and smallest answers that work, and crossing off any answers bigger than the biggest answer that works and smaller than the smallest answer that works.

Question 14 of 20

Last year, Company X spent between $\frac{1}{5}$ and $\frac{1}{4}$ of its yearly advertising budget on print ads. If Company X spent $31,120 on print ads, which of the following could have been Company X's advertising budget last year?

*Indicate **all** such values.*

- ☐ $106,720
- ☐ $115,880
- ☐ $121,960
- ☐ $126,400
- ☐ $138,240
- ☐ $145,000
- ☐ $157,100

Here's How to Crack It

With All That Apply questions, the GRE will often say "which of the following could be." That, and the presence of numbers in our answer choices, tells us that this is a PITA question. Write down A B C D E F G vertically on your scratch paper. Write down the answers and label them as "ad budget." For each answer, we'll calculate $\frac{1}{5}$ and $\frac{1}{4}$ of each ad budget, and see if $31,120 lies between those two numbers.

Let's start with (D), our middle answer. If the ad budget was $126,400, then they must have spent between $\frac{1}{4}$ of $126,400 = $31,600 and $\frac{1}{5}$ of $126,400 = $25,280 on print ads. The problem states that Company *X* spent $31,120 on ads, and $25,280 < $31,120 < $31,600, so $126,400 could have been the amount in Company *X*'s advertising budget. Put a check mark next to (D).

We know that (D) works, but we need to know which other answers work. We'll work our way through the smaller answer choices until they don't work anymore, and then through the bigger answer choices until those don't work. Let's try (C) next. If the ad budget was $121,960, then the company must have spent between $\frac{1}{4}$ of $121,960 = $30,490 and $\frac{1}{5}$ of $121,960 = $24,392 on print ads. Company *X*, however, spent $31,120 on advertising, which is more than that $30,490 allowed by answer (C). Since now our budget is too small, cross off answer (A), (B), and (C).

We're still stuck just with answer (D), so let's try the bigger answer choices. If the ad budget was $138,240, then the company must have spent between $\frac{1}{4}$ of $138,240 = $34,560 and $\frac{1}{5}$ of $138,240 = $27,648 on print ads. Since $27,648 < $31,120 < $34,560, the company's total budget could have been $138,240. Put a check mark next to (E), and let's try a bigger number.

For answer choice (F), if the total advertising budget was $145,000, then the company spent between $29,000 and $36,250 on print ads, which matches our answer. Put a check mark next to (F).

We've only got one answer choice left. Remember, with All That Apply questions you must click on every answer that applies, which sometimes means checking many different answers. If the company had $157,100 in its advertising budget, then it spent between $31,420 and $39,275 on print ads. That's too big of a budget, so we can cross off answer (G).

The answers are therefore (D), (E), and (F).

PITA Works, Too

Believe it or not, there will be times when you can plug in some of the information that they give you in order to solve a problem.

Question 11 of 20

On Monday, Dave's Car Emporium had h used cars in his inventory. Dave moved $\frac{1}{4}$ of the cars to his other dealership on Tuesday, and on Wednesday 7 people brought in their used cars as trade-ins. This brought the total number of used cars on the lot to 67.

Quantity A	**Quantity B**
h	84

- ○ Quantity A is greater.
- ○ Quantity B is greater.
- ○ The two quantities are equal.
- ○ The relationship cannot be determined from the information given.

Here's How to Crack It

Is it possible that $h = 71$? Find out by plugging 84 into the problem: If Dave's had 84 cars on Monday and moved $\frac{1}{4}$ of them off the lot, then he would have had 63 cars left. (Because $\frac{1}{4} \times 84 = 21$, and $84 - 21 = 63$.) If he got 7 more on Tuesday, this would have brought his total up to 70. This doesn't match the 67 figure in the question; in fact, it's too big. Therefore, the number h must be less than 84, and the answer is (B).

Quant Comp PITA Quick Quiz

Question 1 of 3

Mendel purchased three pairs of dress pants at a men's store. One pair cost $65, a second pair cost $85, and the average price for all three was $73.

Quantity A	Quantity B
The price of the third pair of pants	$72

- ○ Quantity A is greater.
- ○ Quantity B is greater.
- ○ The two quantities are equal.
- ○ The relationship cannot be determined from the information given.

Question 2 of 3

A rectangular swimming pool with a uniform depth is 20 feet long and 40 feet wide and it holds a total of 9,600 cubic feet of water.

Quantity A	Quantity B
The depth of the swimming pool	12 feet

- ○ Quantity A is greater.
- ○ Quantity B is greater.
- ○ The two quantities are equal.
- ○ The relationship cannot be determined from the information given.

Question 3 of 3

A local movie theater normally sells individual tickets for $9 each, but it offers 20% discounts to all members of groups of 10 or more.

Quantity A	Quantity B
The total money that a 16-person group spent at the movies	$112

○ Quantity A is greater.

○ Quantity B is greater.

○ The two quantities are equal.

○ The relationship cannot be determined from the information given.

Answers to Quant Comp PITA Quick Quiz

1. Use PITA and assume that the third pair of pants cost $72. If that were the case, then the total cost of all three pairs of pants would be 65 + 85 + 72, or 222, and the average price would be 222 ÷ 3, or $74. Because the problem tells you the average price was $73, the average you calculated is too large, which means the $72 that you plugged into the problem is too large. Therefore, $72 must be greater than the price of the third pair of pants, and the answer is (B).
2. If the dimensions of the swimming pool are 20 by 40 by 12, the volume of the pool is 20 × 40 × 12, or 9,600 cubic feet. Because this matches the volume mentioned in the question, the answer is (C).
3. First, it's important to recognize that the group is large enough to qualify for the discount. If the group spent $112 at the movies, then each spent $112 ÷ 16, or $7. This is too low, because the discounted price of a ticket is 80% of $9, or $7.20. Therefore, they must have spent more than $112, and the answer is (A).

We've run through a number of examples in this chapter, and now it's your turn. As you practice these questions, train yourself to recognize patterns and determine the numbers that will either undermine or confirm your initial results.

Plug In Early, Plug In Often

If you finish this book and take away just one new skill, make this the one. Plugging In is a million-dollar idea that can bring about solid score improvements right away, because of how efficiently it (1) makes problems more accessible and (2) reduces the chance that you'll make careless errors. You should practice Plugging In as much as you can until it becomes instinctive.

Algebra Drill

Question 1 of 35

$ac < bc$

Quantity A	**Quantity B**
a	b

○ Quantity A is greater.

○ Quantity B is greater.

○ The two quantities are equal.

○ The relationship cannot be determined from the information given.

Question 2 of 35

A video game goes on sale for 15% off the original price. A customer can save an additional 10% on the sale price if she has a coupon. If Bailey buys the video game at the sale price and uses a coupon, what percent of the original price does she pay?

○ 12.5%

○ 25%

○ 75%

○ 76.5%

○ 83.5%

Question 3 of 35

A merchant sold an equal number of 5-cent and 10-cent screws. If the total cost of the screws was $3.00, what was the total number of screws sold?

○ 25

○ 30

○ 40

○ 44

○ 50

Question 4 of 35

At the rate of 12,000 rotations per hour, how many rotations will a spinning top make in p minutes?

○ $12{,}000p$

○ $200p$

○ $\frac{72{,}000}{p}$

○ $\frac{12{,}000}{p}$

○ $\frac{200}{p}$

Question 5 of 35

If $3x = 12$, then $8 \div x =$

[ ]

Click on the answer box, then type in a number. Backspace to erase.

Question 6 of 35

$st = -6$

Quantity A	**Quantity B**
s	t

○ Quantity A is greater.

○ Quantity B is greater.

○ The two quantities are equal.

○ The relationship cannot be determined from the information given.

Question 7 of 35

A contest winner received $\frac{1}{4}$ of his winnings in cash and received four prizes, each worth $\frac{1}{4}$ of the balance. If the cash and one of the prizes were worth a combined total of $35,000, what was the total value of his winnings?

- ○ $70,000
- ○ $75,000
- ○ $80,000
- ○ $95,000
- ○ $140,000

Question 8 of 35

Connie has more marbles than Joey has, and Joey has fewer marbles than Mark has.

Quantity A	**Quantity B**
The number of marbles Joey has	The number of marbles Mark has

- ○ Quantity A is greater.
- ○ Quantity B is greater.
- ○ The two quantities are equal.
- ○ The relationship cannot be determined from the information given.

Question 9 of 35

If $2 < r < 8$ and $1 < s < \frac{5}{2}$, which of the following expresses all possible values of rs ?

- ○ $1 < rs < 5$
- ○ $2 < rs < 20$
- ○ $\frac{5}{2} < rs < 8$
- ○ $\frac{5}{2} < rs < 20$
- ○ $5 < rs < 10$

Question 10 of 35

Set A $\{-1, 0, 1\}$

Set B $\{-2, -1, 0, 1\}$

If a is a member of Set A and b is a member of Set B, what is the least possible value of $a - b^2$?

- ○ 2
- ○ 0
- ○ –2
- ○ –5
- ○ –9

Question 11 of 35

$12m^2 - 8m - 64 =$

- ○ $4(3m + 8)(m - 2)$
- ○ $4(3m - 8)(m + 2)$
- ○ $4(3m - 2)(m + 8)$
- ○ $4m^2 - 64$
- ○ $4m - 64$

Question 12 of 35

If $a + b - 2c = 12$, and $3a + 3b + c = 22$, how much is c ?

- ○ –2
- ○ 0
- ○ 10
- ○ 17
- ○ 34

Question 13 of 35

Kevin decided to consecutively number the t-shirts in his closet. He wrote one number on each of his t-shirts, starting with 1 on the first t-shirt. When he was finished numbering he had written a total of 59 digits.

Quantity A	**Quantity B**
35	The number of t-shirts in Kevin's closet

- ○ Quantity A is greater.
- ○ Quantity B is greater.
- ○ The two quantities are equal.
- ○ The relationship cannot be determined from the information given.

Question 14 of 35

The sum of x distinct positive integers is less than 75. What is the greatest possible value of x ?

- ○ 8
- ○ 9
- ○ 10
- ○ 11
- ○ 12

Question 15 of 35

An office supply store sells staplers for $5 each, and boxes of staples for $2 each. On Monday, the store sold a total of 22 staplers and boxes of staples combined, for which it collected a total of $74. How many staplers did the store sell?

- ○ 6
- ○ 10
- ○ 11
- ○ 12
- ○ 44

Question 16 of 35

Carmen has 12 tchotchkes, and t, the value of Carmen's favorite tchotchke, is between 8 and 9 dollars.

Quantity A	**Quantity B**
$\dfrac{t}{\text{the number of tchotchkes Carmen has}}$	0.72

- ○ Quantity A is greater.
- ○ Quantity B is greater.
- ○ The two quantities are equal.
- ○ The relationship cannot be determined from the information given.

Question 17 of 35

Which of the following is equivalent to $8a + (2ab - 4a)b - 4ab$?

- ○ $2a(b - 2)^2$
- ○ $2a(b^2 - 2b + 2)$
- ○ $4a(b^2 - 2b + b)$
- ○ $4ab(1 - 2b)$
- ○ $a^2(2b + 2b^2)b$

Question 18 of 35

Timmie can buy his favorite pens in $10 packs that contain p pens, or he can buy the same pens singly at a cost of $1.12 each.

Quantity A	**Quantity B**
9	The largest possible value of p if it is cheaper to buy the pens singly rather than in packs

- ○ Quantity A is greater.
- ○ Quantity B is greater.
- ○ The two quantities are equal.
- ○ The relationship cannot be determined from the information given.

Question 19 of 35

If $x = \frac{9y}{4}$ and $x \neq 0$, then $\frac{6y}{4x} =$

○ $\frac{27}{8}$

○ $\frac{9}{4}$

○ $\frac{3}{2}$

○ $\frac{2}{3}$

○ $\frac{1}{9}$

Question 20 of 35

If x is an integer less than or equal to –2 and y is an integer whose absolute value is greater than or equal to 5, which of the following must be true of xy ?

*Select **all** that apply.*

☐ xy is an integer

☐ xy is negative

☐ $xy \leq -10$

☐ $xy \geq 10$

☐ $xy = -10$

☐ $|xy| \geq 10$

Click on your choice(s).

Question 21 of 35

Six years ago, Jim's age was four times Carol's.

Jim is now j years old, and Carol is now c years old.

Quantity A	**Quantity B**
c	$\frac{j+16}{4}$

○ Quantity A is greater.

○ Quantity B is greater.

○ The two quantities are equal.

○ The relationship cannot be determined from the information given.

Question 22 of 35

Quantity A	**Quantity B**
$\frac{x^4}{x^3}$	x

○ Quantity A is greater.

○ Quantity B is greater.

○ The two quantities are equal.

○ The relationship cannot be determined from the information given.

Question 23 of 35

$$\frac{a+b}{2} = a^2 - b^2 = 32$$

Quantity A	Quantity B
$a + b$	$(a - b)^2$

- ○ Quantity A is greater.
- ○ Quantity B is greater.
- ○ The two quantities are equal.
- ○ The relationship cannot be determined from the information given.

Question 24 of 35

If $a > 12$ and $b < 7$, which of the following must be true?

- ○ $a + b > 12$
- ○ $a - b > 5$
- ○ $a + b < 19$
- ○ $a - b < 12$
- ○ $ab > 84$
- ○ $ab < 84$

Question 25 of 35

$$x + y = 14$$

$$y + 4 = 10$$

Quantity A	Quantity B
x	y

- ○ Quantity A is greater.
- ○ Quantity B is greater.
- ○ The two quantities are equal.
- ○ The relationship cannot be determined from the information given.

Question 26 of 35

If $x \neq 0$, which of the following must be true?

- ○ $x < x^2$
- ○ $\frac{1}{x} < x$
- ○ $x^2 < x^3$
- ○ $1 - x < x$
- ○ $x < x + 2$

Question 27 of 35

If $x > 0$, $y > 0$, and $z > 0$, then $\frac{2}{x} + \frac{y + \frac{1}{z}}{2} =$

- ○ $\frac{2x}{2x + 2y}$
- ○ $\frac{4 + xy + x}{2x}$
- ○ $\frac{y + z}{2x}$
- ○ $\frac{xyz + 4}{x + y + z}$
- ○ $\frac{4z + xyz + x}{2xz}$

Question 28 of 35

$$2x^2 + 3xy - 2y^2 = 0$$

Quantity A	Quantity B
$2x$	y

- ○ Quantity A is greater.
- ○ Quantity B is greater.
- ○ The two quantities are equal.
- ○ The relationship cannot be determined from the information given.

Question 29 of 35

Pat has 4 teapots more than Judi does, but 3 teapots fewer than Rudy does. If Pat has y teapots, how many teapots do Judi and Rudy have combined?

- ○ $2y - 7$
- ○ $2y - 1$
- ○ $2y + 9$
- ○ $y + 7$
- ○ $y + 9$

Question 30 of 35

If $x \neq 0$ and one-half of x is equal to four times x^2, then $x =$

Click on the answer box, then type in a number. Backspace to erase.

Question 31 of 35

$q \neq 0$

Quantity A	**Quantity B**
$\dfrac{\lvert q-7 \rvert}{2}$	$\dfrac{\lvert q \rvert + \lvert -7 \rvert}{2}$

- ○ Quantity A is greater.
- ○ Quantity B is greater.
- ○ The two quantities are equal.
- ○ The relationship cannot be determined from the information given.

Question 32 of 35

For all non-zero numbers x and y, if $3x = 5y$, then $\dfrac{4x^2}{25y^2} =$

Click on the answer box, then type in a number. Backspace to erase.

Question 33 of 35

Terrence has $\dfrac{2}{3}$ as many cards as does Phillip. If Terrence were to win one card from Phillip, he would have $\dfrac{3}{4}$ as many cards as would Phillip. If Terrence wins an even number of cards from Phillip, which of the following could be the number of cards that Phillip has left?

*Indicate **all** possible values.*

- ☐ 24
- ☐ 20
- ☐ 19
- ☐ 18
- ☐ 17
- ☐ 14

Click on your choice(s).

Question 34 of 35

$$40 < t < 50$$

$$10 < s < 12$$

Quantity A	Quantity B
$t - s$	39

- ○ Quantity A is greater.
- ○ Quantity B is greater.
- ○ The two quantities are equal.
- ○ The relationship cannot be determined from the information given.

Question 35 of 35

If x and n are integers and $x^{(n+2)} = 16x^2$, what is the least possible value of $16 + x$?

[ ]

Click on the answer box, then type in a number. Backspace to erase.

EXPLANATIONS FOR ALGEBRA DRILL

1. D

Plug in. Let $a = 2$, $b = 5$, and $c = 1$. Quantity B can be greater than Quantity A, so eliminate answer choices (A) and (C). Now change the sign of c. Let $a = 5$, $b = 2$, and $c = -1$. Quantity A can be greater than Quantity B, so eliminate answer choice (B). Only choice (D) remains.

2. D

Plug In. $100 for the original price. The sale price is $100 – $15 = $85. The coupon reduces the price another $8.50, to $76.50. This is 76.5% of the original price, making Choice (D) correct.

3. C

Plug In the answer choices, starting with choice (C). If there are 40 total screws, then there are 20 5-cent screws and 20 10-cent screws. The 5-cent screws cost 20 × $.05 = $1.00, and the 10-cent screws cost 20 × $.10 = $2.00. The total cost of the screws is $3.00. This information matches all of the information given in the problem, so choice (C) is correct.

4. B

Plug In, and let $p = 30$ minutes. Because the top rotates 12,000 times per hour, it must rotate 6,000 times in 30 minutes, which is one half-hour. Plug 30 for p into the answer choices to find the target answer of 6,000. You'll find it only in choice (B): 200 × 30 = 6,000.

5. 2

First, solve for x by dividing both sides of the equation by 3. $x = 4$. Next, answer the question: $8 \div 4 = 2$.

6. D

There are variables in the columns, so Plug In. Try $s = 2$ and $t = -3$. Quantity A is bigger, so cross off choices (B) and (C). Now switch it and try $s = -3$ and $t = 2$. This time Quantity B is bigger, so the correct answer is choice (D).

7. C

Plug In the answers, starting with choice (C). If the winnings were $80,000, the cash is $\frac{1}{4} \times \$80{,}000 = \$20{,}000$. Then, take $\frac{1}{4}$ of the remainder of the prize, $60,000: $\frac{1}{4} \times \$60{,}000 = \$15{,}000$. The total of the cash and the first payment is $20,000 + $15,000 = $35,000. Because choice (C) is correct, there is no need to check the other answers.

8. D

There are unknowns in the columns, so set up your scratch paper and Plug In more than once. Try 10 marbles for Connie, 5 marbles for Joey, and 7 marbles for Mark. Quantity B is bigger, so cross off choices (A) and (C). Now try 10 marbles for Connie, 5 marbles for Joey, and 10 marbles for Mark. This time the quantities are equal, and you can eliminate choice (B). The correct answer is choice (D).

9. B

Break this problem into two parts, and use POE. Because $r > 2$ and $s > 1$, the product of rs must be greater than the product of 2×1. The only answer choice in which $2 < rs$ is choice (B). In addition, because $r < 8$ and $s \frac{5}{2}$, the product of rs must be less than $8\left(\frac{5}{2}\right) = 20$. Choice (B) is also the only answer choice in which $rs < 20$.

10. D

To make the value of $a - b^2$ as small as possible, make a as small as possible and b as large as possible. The smallest number is Set A is -1, so that's the value for a. Use -2 for b because that makes $b^2 = 4$: $a - b^2 = -1 - 4 = -5$.

11. B

Plug In, and let $m = 3$. The equation is $(12 \times 3^2) - (8 \times 3) - 64 = 108 - 24 - 64 = 20$, and 20 is the target. Plug 3 for m into the answer choices, and match the target of 20. Only choice (B) matches the target: $4(3 \times 3 - 8)(3 + 2) = 4(9 - 8)(5) = 4(1)(5) = 20$.

12. A

Since you're looking for c, try to make a and b cancel out of the equations. Multiply the first equation by -3, giving you $-3a - 3b + 6c = -36$. Now add the equations. The a and b values cancel out, giving you $7c = -14$, so $c = -2$.

13. A

There is an unknown in one column and a number in the other, so PITA. If there were 35 t-shirts in Kevin's closet that would be 9 digits for the first 9 t-shirts and $2 \times 26 = 52$ for the next 26. This a total of 61 digits, which is more than 59 mentioned in the question, so the number of t-shirts must have been less than 35. The correct answer is choice (A).

14. D

Write down the integers starting with 1 and keep adding: $1 + 2 + 3 + 4 + 5 + 6 + 7 + 8 + 9 + 10 + 11 = 66$. If you added 12 to the total, you would reach 78. Thus, there are 11 distinct positive integers that have a sum less than 75.

15. B

Label the number of staplers sold x, and the number of boxes of staples sold y; now you can write the equation $x + y = 22$. Additionally, you know that $\$5x + \$2y = \$74$. You have two equations and two unknowns, so stack the equations in order to add or subtract them. Since you're solving for x, multiply the first equation by 2 to get $2x + 2y = 44$; now you have the same number of y's, so they'll cancel out if you subtract one equation from the other. Subtracting gives you $3x = 30$, so $x = 10$.

16. D

Once you've set up your scratch paper to Plug In on a Quant Comp problem, start with $t = 8.1$. The value in Quantity A is 0.675; Quantity B is greater, so eliminate choices (A) and (C). Next, try $t = 8.9$. Now the value in Quantity A is about 0.74; this time, Quantity A is greater, so eliminate choice (B) and select choice (D).

17. A

Plug In. If $a = 2$ and $b = 3$, then $8a + (2ab - 4a)b - 4ab = 8(2) + (2(2)(3) - 4(2))(3) - 4(2)(3) = 16 + (12 - 8)(3) - 24 = 16 + 12 - 24 = 4$. Your target is 4. Plug $a = 2$ and $b = 3$ into all of the answers. Only answer choice (A) yields 4: $2(2)(3 - 2)^2 = 4(1) = 4$.

18. A

There's an unknown in one column and a number in the other, so Plug In the Answers. If p is 9, Timmie pays $9 \times \$1.12 = \$10.08$ to buy the pens singly; that's more than it would cost him to buy the pens in the pack, but it's supposed to be cheaper to buy the pens singly. Therefore, p must be less than 9, and the correct answer is choice (A).

19. D

Plug in. If $y = 4$, $x = \dfrac{(9)(4)}{4} = 9$. The problem is asks for $\dfrac{6y}{4x}$, which is $\dfrac{(6)(4)}{(4)(9)} = \dfrac{6}{9} = \dfrac{2}{3}$. Choice (D) is the only match.

20. A and F

The product of two integers must be an integer, so answer choice (A) is correct. The requirement that $|y| \geq 5$ is equivalent to the following conditions: $y \geq 5$ or $y \leq -5$. If $y \geq 5$ and $x \leq -2$, then $xy \leq -10$; if $y \leq -5$ and $x \leq -2$, then $xy \geq 10$. Hence, $xy \leq -10$ or $xy \geq 10$, which is the same thing as saying $|xy| \geq 10$, so choice (F) is correct. Answer choice (B) is only true of some values of xy; choices (C) and (D) are partial answers along the way to choice (F), but are incomplete and hence incorrect. Of course, you can simplify this problem greatly by plugging in for x and y. First, make $x = -2$ and $y = 5$: $xy = -10$, so eliminate answer choice (D). Next, leave $x = -2$ but make $y = -5$: Now $xy = 10$, so eliminate answer choices (B), (C), and (E). No matter what variable you plug in—as long, naturally, as you meet the requirements—answer choices (A) and (F) will always work, so they *must* be true.

21. A

Plug in a few different rounds of numbers. The question states $j - 6 = 4(c - 6)$. If you choose $c = 7$, then $j = 10$. In that case, $7 > 6.5$ and Quantity A is greater, so you can eliminate choices (B) and (C). To decide between answers (A) and (D), choose a couple of very different numbers to plug in, but recall that negative numbers and zeros don't work for ages. Let $c = 66$ to check the high end of the range. For $c = 66$, $j = 246$, and $66 > 65.5$, and Quantity A is still 0.5 greater. Choose one more number to be sure. If $c = 10$, $j = 22$. Once again Quantity A is 0.5 greater than Quantity B because $10 > 9.5$, so you can confidently select choice (A).

22. C

If you remember your exponent rules, you know that Quantity A reduces to x. However, you should always plug in on QC questions, and when you're dealing with exponents or absolute values, you must remember to plug in negative numbers. Go through ZONEF. If $x = 0$, then $0 = 0$ and the expressions are equal: Eliminate choices (A) and (B). If $x = 1$, then $1 = 1$. If $x = -2$, then $-2 = -2$. If $x = 100$, then $100 = 100$. If $x = \frac{1}{2}, \frac{1}{2} = \frac{1}{2}$. At this point you can be confident that choice (C) is the best answer.

23. A

Recognize the common quadratic at play and this becomes easier. You are initially given $a + b$ and $a^2 - b^2$, and this may remind you of the common quadratic $(a + b)(a - b) = a^2 - b^2$. From the other given information, solve for $a + b$, which equals 16. Substituting 16 for $(a + b)$ and 32 for $(a^2 - b^2)$ in the quadratic, you can solve for $(a - b) = 2$. Now Quantity A is 16 and Quantity B is $2^2 = 4$, making Quantity A greater.

24. B

Solve this must be problem by plugging in. Let $a = 13$ and $b = 6$. These numbers eliminate choices (C) and (E). Then try some extremes, such as $a = 100$ and $b = -100$. These new numbers eliminate choices (A) and (D). Try $a = 100$ and $b = 1$, which eliminates choice (F). Only choice (B) always works, because the values of a and b must be more than 5 apart.

25. A

Solve the second equation and you'll find that $y = 6$. Plug that in to the first equation, and you'll find that $x = 8$.

Therefore, Quantity A is greater.

26. E

Plug In different numbers until there is only one answer choice left. If you try a simple number such as $x = 2$, you'll find that you can't eliminate any answer choices. That means it's time to start thinking about different kinds of numbers. If $x = \frac{1}{2}$, choices (A), (B), (C), and (D) are eliminated. Only choice (E) works.

27. E

Plug In $x = 2$, $y = 3$, and $z = 4$. Put these values into the equation: $\frac{2}{2} + \frac{3+\frac{1}{4}}{2} = \frac{2+3+\frac{1}{4}}{2} = \frac{5\frac{1}{4}}{2} = \frac{\frac{21}{4}}{2} = \frac{21}{4} \times \frac{1}{2}$ $= \frac{21}{8}$. Now go through the answer choices and find $\frac{21}{8}$. The only one that works is answer choice (E).

28. D

You can effectively plug-in answer choice (C) by assuming for a moment that $2x = y$. When you substitute this into the quadratic equation, it satisfies it and validates choice (C). Eliminate choices (A) and (B), neither of which can be the final answer if choice (C) is ever correct. Factor the quadratic equation: $(2x - y)(x + 2y) = 0$. This means $2x = y$ (which you proved above) or $x = -2y$. Plug in to this second scenario, which you have yet to examine. When $x = 0$, then $y = 0$ and the columns are still equal, validating choice (C). But when $x = 1$, $y = -.5$, making Quantity (A) bigger. So choice (D) must be the correct answer.

29. B

Plug In. If $y = 6$, then Pat has 6 teapots, Judi has 2 teapots, and Rudy has 9 teapots. The target answer, the number of teapots that Judi and Rudy have combined, is 11. Go to the answer choices, and Plug In 6 for y. Answer choice (B) is the only answer choice that matches your target of 11.

30. $\frac{1}{8}$

Translate the words in the question into an equation: $\frac{1}{2}x = 4x^2$. Because the question states that $x \neq 0$, you can divide both sides of the equation by x to get $\frac{1}{2} = 4x$. Divide both sides by 4 to find that $x = \frac{1}{8}$.

31. D

Notice that the trap answer is choice (C). There are variables in the columns, so Plug In. Try $q = 7$. Quantity A is 0 and Quantity B is 7. Quantity B is bigger, so cross off choices (A) and (C). Now try $q = -7$. This time both columns are 7. That's two different answers, so the correct answer is choice (D).

32. $\frac{4}{9}$

Plugging In is the easiest way to solve this question. Let $x = 5$ and $y = 3$. Plug those numbers into the expression:

$$\frac{4x^2}{25y^2} = \frac{4(5)^2}{25(3)^2} = \frac{(4)(25)}{(25)(9)} = \frac{4}{9}.$$

33. C and E

The first sentence tells you that $T = \frac{2}{3}P$, and the second tells you that $(T + 1) = \frac{3}{4}(P - 1)$. Simplify the second equation to get: $T = \frac{3}{4}P - \frac{7}{4}$. Because both equations express the value of T, set them equal to one another: $\frac{2}{3}P = \frac{3}{4}P - \frac{7}{4}$. Multiply by 12 to get rid of the fractions, and you're left with $8P = 9P - 21$; P, therefore, equals 21. If Phillip has 21 cards and loses an even number of them, he could have $21 - 2 = 19$ cards or $21 - 4 = 17$ cards left. Answer choices (C) and (E) are correct.

34. D

If you simply subtract the second inequality from the first, it can appear as if choice (B) should be the answer. Be careful of merely doing the obvious on a question like this. Because there are variables, set up to plug in on a quant comp. Choose something easy to start with, like $t = 45$ and $s = 11$. $t - s = 45 - 11 = 34$. Eliminate choices (A) and (C). Your task is to now attempt to make the value of column A greater than (or equal to) 39. This appears difficult at first, because of the limited amount of range you're given. ZONEF isn't particularly helpful: You can't plug in 0, 1, or negatives for this problem, but you can try fractions. Let $t = 49$ and $s = 10.5$. $49 - 10.5 = 38.5$. You're closer, but still not quite there. This time, let $t = 49.5$ and keep $s = 10.5$. $49.5 - 10.5 = 39$. Eliminate choice (B), and select choice (D).

35. 12

To find the least possible value of $16 + x$, first find the least possible value of x. Rewrite the equation as $(x^n)(x^2) = 16x^2$. Divide both sides by x^2 to get $x^n = 16$. If both x and n are integers, then there are only a very few values for x and n. Minimize $16 + x$ by using a negative value for x. Because, $(-4)^2 = 16$, $x = -4$ is the least possible value for x. The least possible value of $16 + x$ is $16 - 4 = 12$.

Chapter 5
Charts and Crafts

THE JOY OF VISUAL DATA

GRE chart questions are pretty straightforward; you read a couple of charts and/or graphs and then you answer some questions about them.

Chart questions will appear on a split screen; the chart(s) will appear on the left, and the questions will be on the right. When a chart question pops up, be sure to start by hitting the scroll bar to see just how much data you're dealing with. There might be a second chart hidden below the first one, so don't get thrown off if you see questions about a chart you don't think exists.

Chart questions test four primary skills.

- how well you read charts
- how well you approximate percentage and percentage change
- how well you calculate exact percentage and percentage change
- how well you synthesize related data from two separate sources

When you have to answer chart questions, there are a couple things to remember.

1. **Read all the charts.** Before you answer any questions about a set of charts, look over the charts. Make sure you pay particular attention to the units involved. Is each axis in terms of a number of people, a percentage, a dollar amount? **Look for a legend for the charts.** Are we talking about 100 people, or 100 million people? **Check to see if you need to scroll down for any additional charts.** If there are multiple charts, spend a minute or so figuring out how the charts relate to each other. Do the charts show different aspects of the same information? For instance, does one chart show the jobs of a group of people, whereas the other chart shows the education level of those same people? Does one chart give more detail about a limited sliver of information from the previous chart? Spending time understanding the charts now will pay off later.
2. **Find the information you need.** Once you read a charts question, figure out which chart and which data points you need to look at.
3. **Use your scratch paper and label your information.** Find the data points you need to answer the question, and write each one down on your scratch paper. Include units and a name for each data point. Don't just write "23." Write: "2004 imports – 23 mil tons."It only adds an extra couple of seconds of time, and makes it much less likely that you will make a mistake or get lost in the problem.
4. **Estimate before you calculate.** Look to eliminate answers before doing too many calculations. If 226 of the 603 employees were fired, and the question wants to know what percentage were fired, then do

a quick estimation: $\frac{226}{603} \approx \frac{200}{600} = \frac{2}{6} = \frac{1}{3} \approx 33\%$. Since we rounded 226 down to 200, our answer will probably be a bit larger than 33%. Cross off any answers that are 33% or lower, or much bigger than 40%. If we can't choose an exact answer based on our estimation, then we can go back and calculate afterwards.

Percentage Change Quick Quiz

Remember all the reading you did about percentages and percent change in Chapter 3? Well, here are some questions. Let's see how much skill you've retained:

Question 1 of 5

If Amy's stock portfolio grew from $1,200 to $1,600 in January, by approximately what percent did it increase during the month?

- ○ 25%
- ○ 33%
- ○ 40%
- ○ 67%
- ○ 75%

Question 2 of 5

If she withdrew 30% of her money in February, how much was left in her brokerage account?

- ○ $360
- ○ $480
- ○ $720
- ○ $840
- ○ $1,120

Question 3 of 5

Amy's portfolio increased by 25% of its value in the first two weeks of March but lost 25% in the last two weeks.

Quantity A	**Quantity B**
The amount left in the account at the end of February	The amount of money left in her account at the end of March

○ Quantity A is greater.

○ Quantity B is greater.

○ The two quantities are equal

○ The relationship cannot be determined from the information given.

Question 4 of 5

If her portfolio increased by 200% during the month of April, what was her new balance at the end of April?

[ ]

Click on the answer box, then type in a number. Backspace to erase.

Question 5 of 5

By approximately what percent did Amy's portfolio increase between January 1 and April 30 ?

○ 32%

○ 49%

○ 62%

○ 163%

○ 213%

Explanations for Percentage Quick Quiz

1. If Amy's stock portfolio grew from $1,200 to $1,600 in January, then the change is 1,600 – 1,200, or $400. The percent change is $\frac{400}{1200} \times 100$ or 33%. The answer is (B).
2. If she started with $1,600 and withdrew 30% of her money, then she withdrew $\frac{30}{100} \times 1600$ or $480. She therefore had $1,600 – $480, or $1,120 left. (Note: You can solve this with one fewer step if you realize that by taking out 30%, Amy left 70% in her account; 70% of 1,600 is 1,120.) The answer is (E).
3. The numbers might suggest that her balance didn't change (25% up and 25% down might look like they cancel out), but as we'll see, it did. If Amy's portfolio increased by 25% of its value in the first two weeks of March, the balance rose by $\frac{25}{100} \times 1120$, or $280, to $1,400. Then, if it lost 25% in the last two weeks, the balance fell from $1,400 to $\frac{75}{100} \times 1400$, or $1,050. She had less money at the end of March than at the end of February, so the answer is (A).
4. The wording of this is a little tricky, because if something increases by 200%, it is adding an amount that is double the original value. (In effect, increasing by 200% is the same thing as *tripling*.) Because 200% of $1,050 is $2,100, the new value of the portfolio at the end of April was $1,050 + $2,100, or $3,150.
5. Amy started out with $1,200, and at the end of April she had $3,150. Over the four months, the portfolio gained $3,150 – $1,200, or $1,950, in value. Without even using your calculator, you should be able to see that the new value is more than double the old value but less than triple the old value. This means that the portfolio increased by between 100% and 200%. The answer must therefore be (D), because no other answer choice is even close.

FOCUS MINIMIZES CARELESSNESS

Reading charts is a lot like interpreting points that are graphed on the coordinate plane. It's a process that anyone can learn with practice. The worst thing about GRE charts, however, is that they are often intentionally confusing and/or difficult to read, so it's easy to make careless mistakes under stressful conditions. Above all else, resolve to be calm and systematic when you navigate the data that the GRE throws at you. If you can manage that, the rest should fall into place nicely.

The Formats

Most chart questions come in three formats: bar or line graphs, pie charts, and data tables. Additionally, the test often gives you two separate charts that are somehow related, however remotely.

Data tables are made up entirely of numbers, so all you need to do is crunch the numbers using your calculator. Graphs and charts can be a little trickier because they require you to interpret them visually.

Bar Graphs

On a bar graphs, data points are indicated with rectangular bars that can run either horizontally or vertically. Bar graphs get tricky when one bar contains two or more bits of data that are often colored or patterned differently, like this:

COMPANY *X*'S GROSS PROFITS,
IN MILLIONS, 2001–02

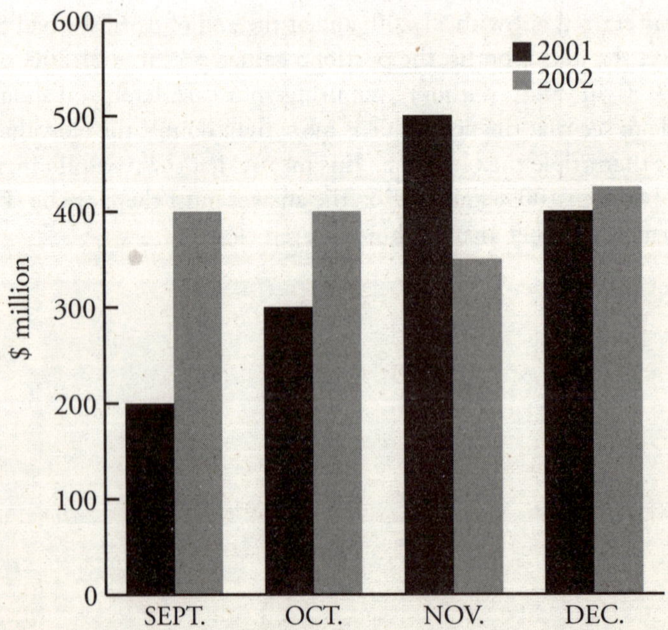

Line Graphs

Line graphs are very similar to bar graphs except they link data points with lines that can indicate overall trends.

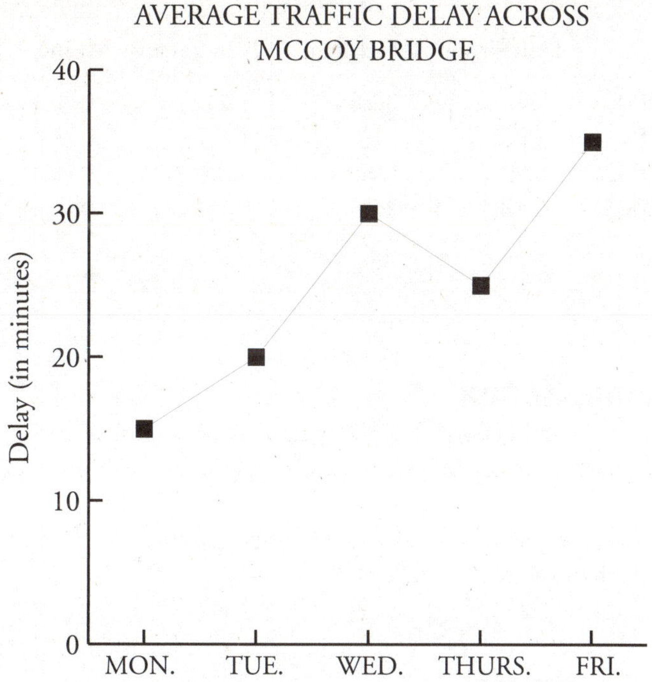

Pie Charts

Pie charts divide up all data into wedges, each of which indicates a percentage of the whole. All of the percentages should total to 100%, and the bigger the wedge, the bigger the percentage.

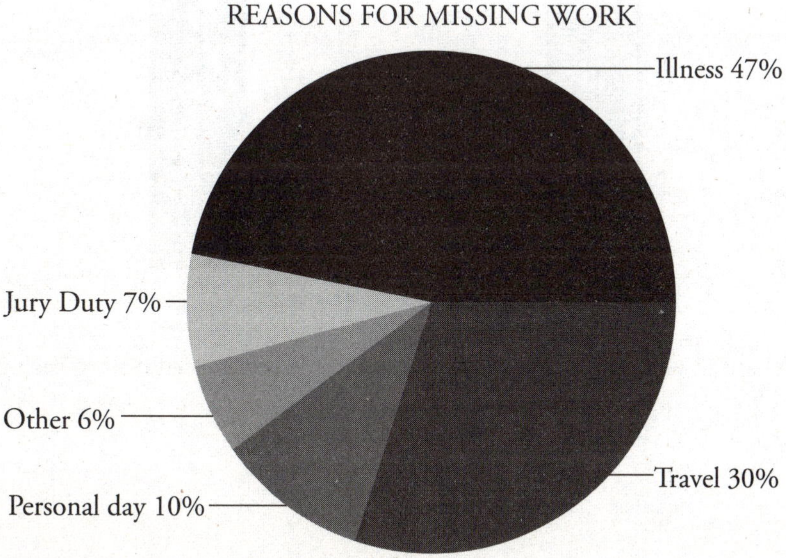

Data Tables

Sometimes the GRE does away with graphics altogether and simply presents raw data in several rows and columns.

Number of Driver's Licenses Applied For, in millions

	California	Michigan	New Jersey	Maine
2000	5.41	2.26	1.21	.061
2001	4.96	1.05	0.95	.039
2002	4.92	1.96	1.06	.044
2003	4.88	2.33	1.15	.052

Ballparking, Redux

Many times, the numbers you'll work with will be approximations rather than exact calculations. Working well with these approximations is an important skill.

Let's look at an example: About how much bigger, in terms of a percentage, is the bar on the left than the bar on the right?

Pretty hard to tell, isn't it? Now look at those same bars placed next to each other:

When the bars are adjacent, it's a little easier to compare their relative size. The bar on the left is about half again as big, or 50% bigger. You can confirm this when the two bars are next to a vertical scale:

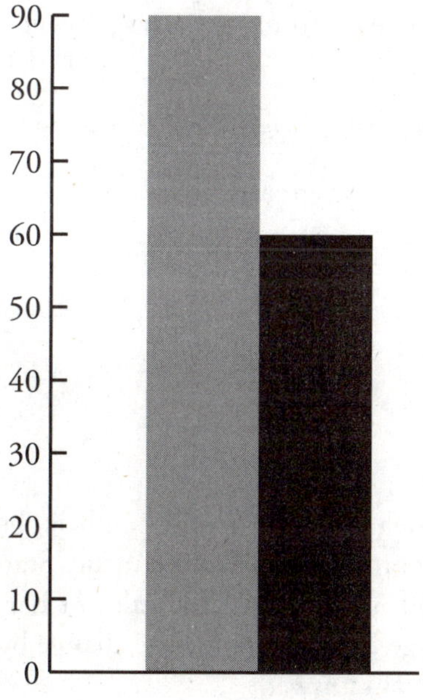

Now you have a very good (though not necessarily exact) idea of how large each bar is, and you can crunch some numbers to determine the percent decrease from Bar A to Bar B or the percent increase from Bar B to Bar A. Let's work that into a sample question, shall we?

Question 16 of 25

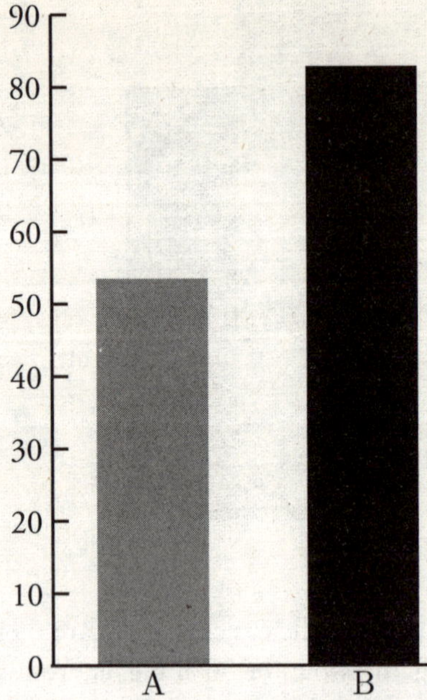

Quantity A	**Quantity B**
The percent increase from Bar A to Bar B	The percent decrease from Bar B to Bar A

- ○ Quantity A is greater.
- ○ Quantity B is greater.
- ○ The two quantities are equal.
- ○ The relationship cannot be determined from the information given.

Here's How to Crack It

For this question, you don't have any actual numbers, but you can plug some of your own in using the approximations you've made. Start by letting Bar A equal 100; if Bar B is around 50% larger, then the value of Bar B can be 150. Now you can use the percent-change formula (divide the change by the original value, then multiply by 100) to find your answer.

The change from A to B is 50, and the original value is 100, so the percent increase is $\frac{50}{100} \times 100$ or 50%. Conversely, the percent decrease from B to A is $\frac{50}{150} \times 100$ or 33%. The answer is (A).

The Power of Guesstimation

Have you ever heard the expression "*Almost* only counts in hand-grenades and horseshoes"? Well, you can add "GRE charts" to that list. Approximating values is an important skill when you're interpreting charts, because as we saw in the previous example, the value might not necessarily sit directly on a calibration line.

If the test asks you to estimate something, the answer choices will not be very close together. If the right answer is 47%, for example, you won't see 44% or 50% among the answer choices because they're too close to the right answer. Someone could easily be just a little off in his estimations and get 44% rather than 47%. ETS doesn't want to leave any room for dispute, because they don't want complaints from test takers.

If you don't feel confident in your abilities to estimate, resolve to practice on as many chart questions as you can until you're more confident. Greater skill will come with time, diligence, and patience.

Ballparking Quick Quiz

What is the percent change between Quantity A and Quantity B in each of the following examples?

Question 1 of 5

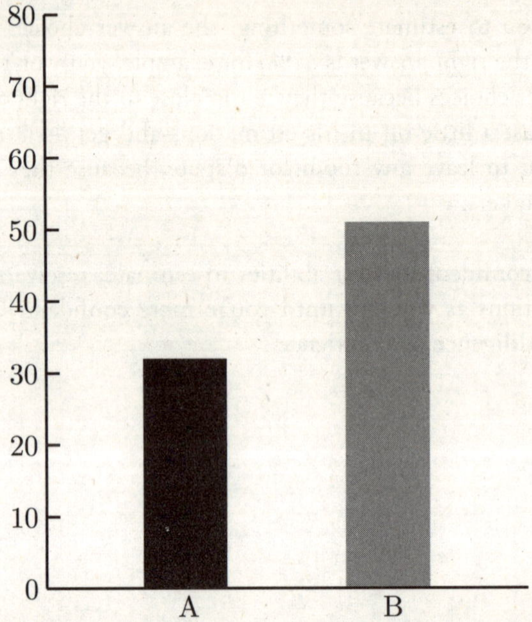

○ 19% increase

○ 25% increase

○ 41% increase

○ 41% decrease

○ 59% increase

Question 2 of 5

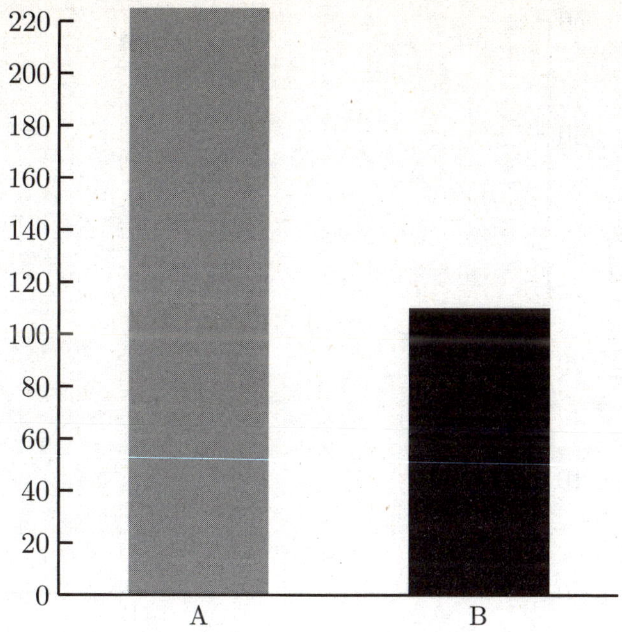

- ○ 25% decrease
- ○ 49% increase
- ○ 51% decrease
- ○ 100% decrease
- ○ 115% decrease

Question 3 of 5

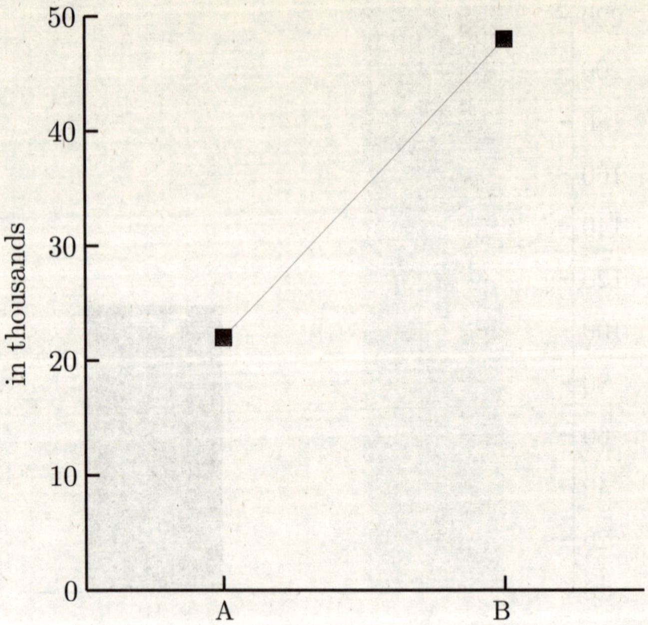

○ 26% decrease

○ 54% decrease

○ 54% increase

○ 118% increase

○ 150% increase

Question 4 of 5

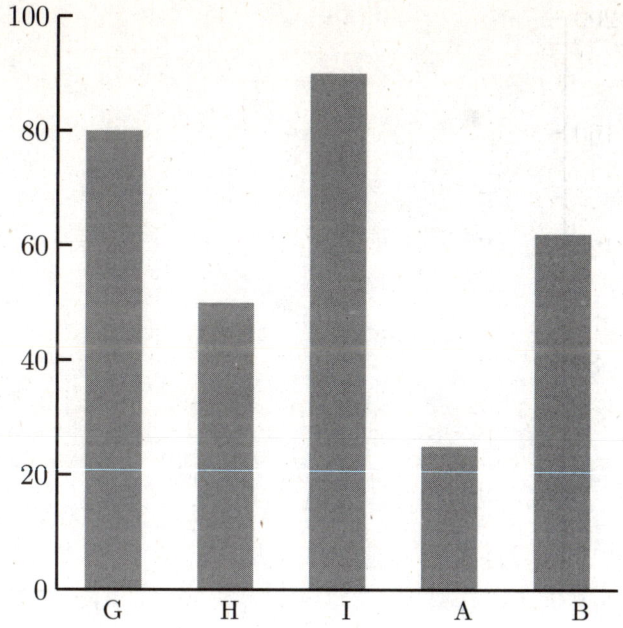

○ 24% increase

○ 48% increase

○ 60% increase

○ 148% increase

○ 248% increase

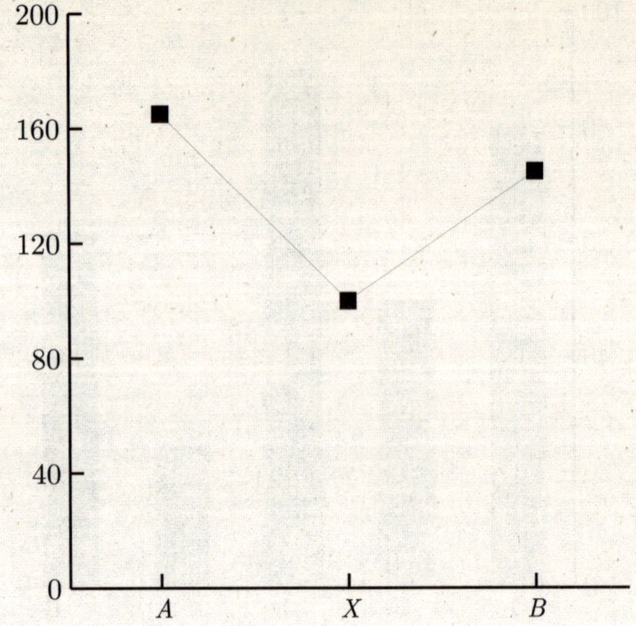

○ 12% decrease

○ 14% increase

○ 20% decrease

○ 24% decrease

○ 38% decrease

Explanations for Ballparking Quick Quiz

To solve each of these five, you have to use the percent-change formula and remember that Bar A is the original value.

1. Bar A has an approximate value of 32, and Bar B's approximate value is 51; the difference is 51 – 32, or 19. The percent change is $\frac{19}{32} \times 100$ or a 59% increase. The answer is (E).
2. Bar A has an approximate value of 225, and Bar B's approximate value is 110; the difference is 225 – 110, or 115. The percent change is $\frac{115}{225} \times 100$, or a 51% decrease. The answer is (C).
3. Value A is approximately 22,000, and Value B is around 48,000; the difference is 48,000 – 22,000, or 26,000. The percent change is $\frac{26000}{22000} \times 100$, or a 118% increase. The answer is (D).
4. This might have seemed a little more difficult, because Bar A and Bar B are farther apart and farther from the left-side measurement. Bar A has an approximate value of 25, and Bar B's approximate value is 62; the difference is 62 – 25, or 37. The percent change is $\frac{37}{25} \times 100$, or a 148% increase. The answer is (D).
5. This might also have been a bit more difficult because of the distractor value in between Values A and B. Value A is approximately 165, and B's approximate value is 145; the difference is 165 – 145, or 20. The percent change is $\frac{20}{165} \times 100$, or a 12% decrease. The answer is (A).

Here's an example of how a series of chart questions might look.

Questions 6–8 refer to the following graphs.

PRESIDENT'S FAVORABLE AND UNFAVORABLE RATINGS, 2002

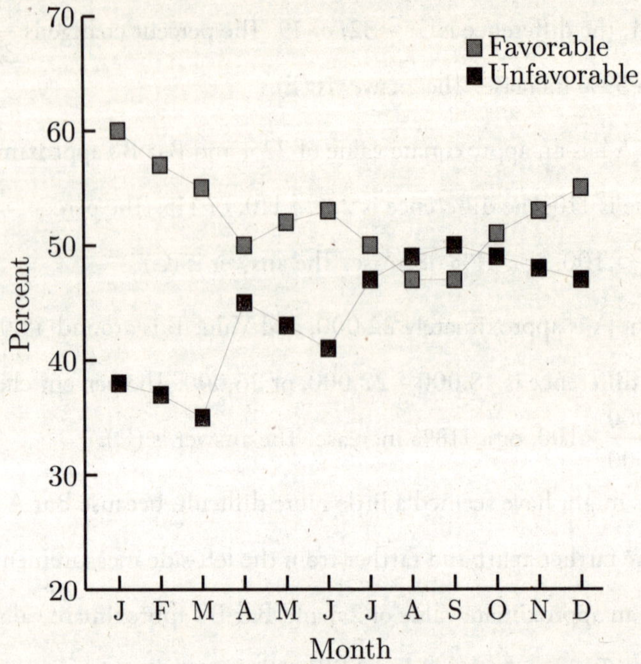

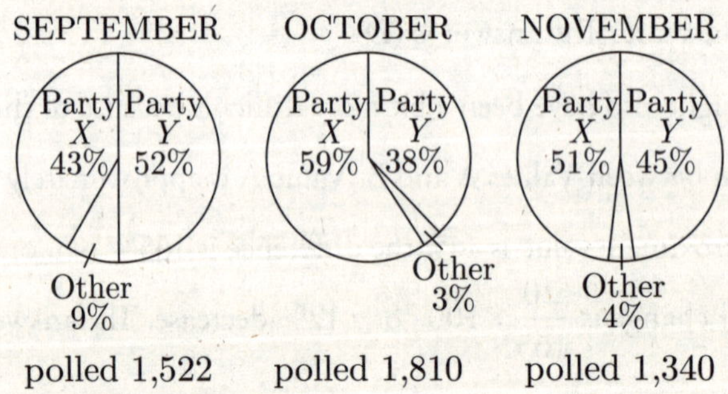

Question 6 of 20

In which month did the difference between the percentages of the favorable and unfavorable ratings of the president shrink to less than 10% ?

○ March

○ April

○ May

○ August

○ November

Here's How to Crack It

The favorable/unfavorable ratings are displayed in the line chart. During the first several months, the gap between the favorable and unfavorable ratings narrows steadily. In January, the favorable ranking is just under 60% and the unfavorable is under 40%. That's too wide. In April, however, the favorable rating drops to around 52%, while the unfavorable rating rises to around 45%. This is clearly less than 10%, and the answer is (B).

Question 7 of 20

How many people who were polled in October identified themselves as members of party Y ?

- ○ 413
- ○ 536
- ○ 603
- ○ 723
- ○ 891

Here's How to Crack It

Now we're looking at the pie charts, because they're the ones that reference the party affiliations of those polled. The November pie is on the far right, and 45% of the 1,340 people said they belonged to party *Y*. Because 1,340 × 0.45 = 603, the answer is (C).

Question 8 of 20

In which month was the greatest percentage of poll respondents undecided?

- ○ March
- ○ May
- ○ June
- ○ October
- ○ November

Here's How to Crack It

We're back to the line graphs, which indicate the percentages of favorable and unfavorable votes. Throughout the chart, the two percentages don't add to 100%; the chart explains that the remaining percentages were undecided. In March, the favorable percentage was around 56%, and the unfavorable percentage was only around 36%. These numbers therefore account for only 92% of the respondents; therefore, 8% of them must have indicated that they were undecided. This is by far the largest value of undecideds among the five answer choices given. The answer is (A).

And that's the basic idea behind Charts questions. The wall of data is supposed to be intimidating, especially under test-taking pressure. But if you pay attention, keep your cool, and work methodically—and keep practicing—you'll find they're really a snap.

Charts and Crafts Drill

Questions 1–5 refer to the following graphs.

EXPENDITURES ON METAL BY COMPANY *X*

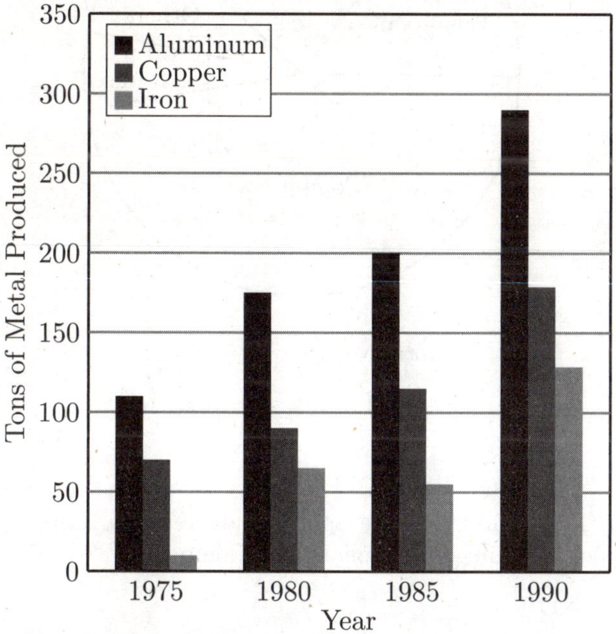

DISTRIBUTION OF SPENDING ON METALS BY COMPANY *X* IN 1990

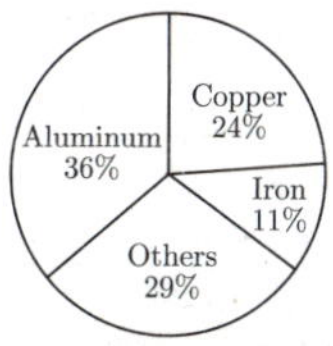

Year	Price of Aluminum Per Ton
1975	\$1,900
1980	\$2,200
1985	\$2,700
1990	\$3,400

Question 1 of 15

Approximately how many tons of aluminum and copper combined were purchased in 1985 ?

- ○ 75
- ○ 265
- ○ 325
- ○ 375
- ○ 470

Question 2 of 15

How much did Company *X* spend on aluminum in 1980 ?

- ○ \$675,000
- ○ \$385,000
- ○ \$330,000
- ○ \$165,000
- ○ \$139,000

Question 3 of 15

Approximately what was the percent increase in the price of aluminum from 1975 to 1985 ?

- ○ 8%
- ○ 16%
- ○ 23%
- ○ 30%
- ○ 42%

Question 4 of 15

In 1990, if Company *X* spent \$3,183,000 on metals, what was the approximate price per ton of iron?

- ○ \$1,040
- ○ \$2,000
- ○ \$2,800
- ○ \$3,400
- ○ \$4,670

Question 5 of 15

Which of the following can be inferred from the graphs?

*Select **all** that apply.*

☐ Company X spent more on copper than it did on iron in 1975.

☐ The price per ton of copper was more than that of aluminum in 1990.

☐ Company X purchased less iron in 1990 than it did "Other" metals in 1990.

Questions 6–10 refer to the following graphs.

PRODUCT SALES AND EMPLOYEES OF FOUR LEADING ELECTRONIC COMPANIES IN 1988

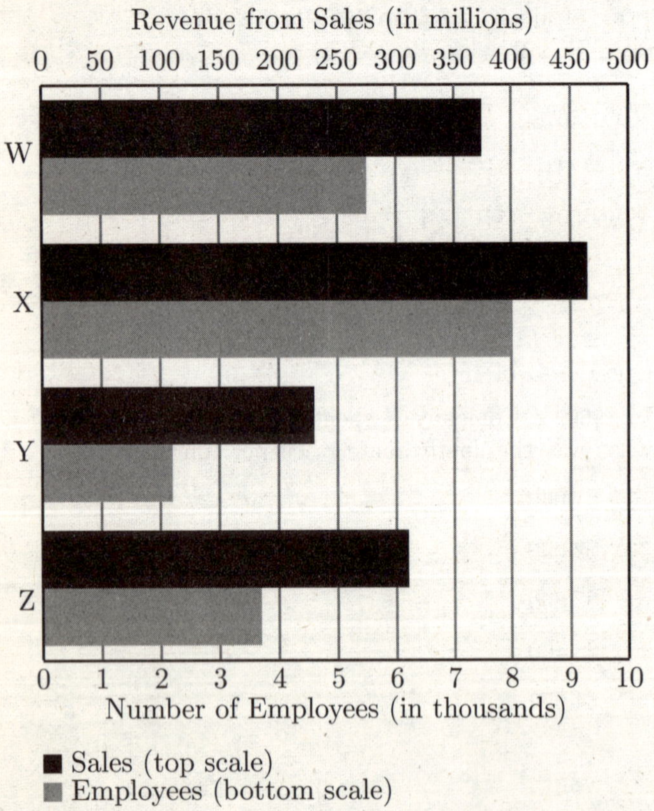

DISTRIBUTION OF SALES RECEIPTS BY PRODUCT COMPANY W, 1988 (in millions of dollars)

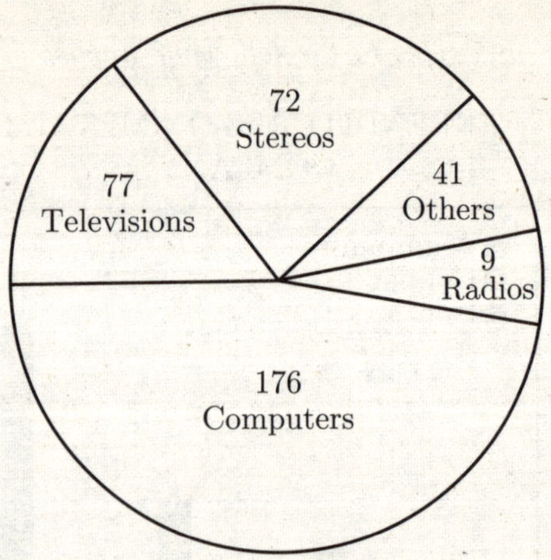

Note: Not drawn to scale.

Question 6 of 15

In 1988, what was the approximate revenue from sales, in millions of dollars, for Company Y?

○ 110

○ 235

○ 310

○ 375

○ 475

Question 7 of 15

In 1988, how many of the four leading electronics companies employed at least 4,000 people?

○ None

○ One

○ Two

○ Three

○ Four

Question 8 of 15

In 1988, which of the four leading electronics companies had the highest revenue from sales per employee?

- ○ *W*
- ○ *X*
- ○ *Y*
- ○ *Z*
- ○ It cannot be determined from the information given.

Question 9 of 15

What was the approximate average (arithmetic mean) of revenue from sales, in millions of dollars, for the four leading electronics companies in 1988 ?

- ○ 290
- ○ 345
- ○ 375
- ○ 420
- ○ 460

Question 10 of 15

By what percentage were receipts from computer sales greater than receipts from stereo sales for Company *W* in 1988 ?

- ○ 7%
- ○ 59%
- ○ 104%
- ○ 144%
- ○ 244%

Questions 11–15 refer to the following graphs.

EXPORTS BY SCANDINAVIAN COUNTRIES IN YEAR *X*

Total — $270 billion

By Country

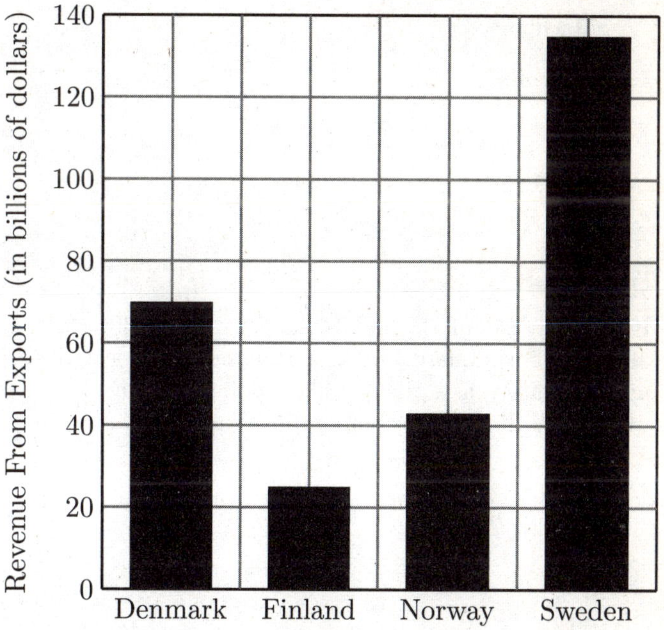

By Product

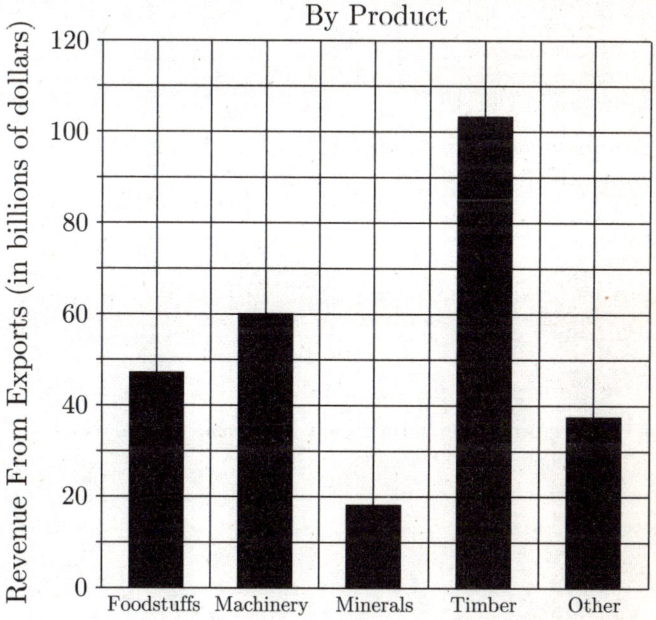

Question 11 of 15

In Year X, which of the following products represented between 15 percent and 20 percent of the total revenue from all Scandinavian exports?

- ○ Foodstuffs
- ○ Machinery
- ○ Minerals
- ○ Timber
- ○ Other

Question 12 of 15

If Sweden, Denmark, and Finland exported 30, 18, and 6 billion dollars worth of machinery, respectively, approximately how many billion dollars worth of machinery did Norway export in Year X ?

- ○ 62
- ○ 43
- ○ 19
- ○ 12
- ○ 8

Question 13 of 15

Approximately how many billions of dollars worth of minerals did Denmark export in Year X ?

- ○ 5
- ○ 18
- ○ 48
- ○ 67
- ○ It cannot be determined from the information given.

Question 14 of 15

In Year X, if foodstuffs accounted for 15 percent of the revenue from all Swedish exports, and if the revenue from foodstuffs exported by Norway was 40 percent less than the revenue from foodstuffs exported by Sweden, then foodstuffs accounted for approximately what percent of the revenue from all Norwegian exports?

- ○ 8%
- ○ 20%
- ○ 28%
- ○ 44%
- ○ 60%

Question 15 of 15

If, on average, 10 pounds of exported timber were worth $1, and 4 pounds of exported foodstuffs were worth $1, what was the approximate ratio of the number of pounds of exported timber to the number of pounds of exported foodstuffs for the four Scandinavian countries in Year X ?

- ○ 5 : 4
- ○ 2 : 1
- ○ 5 : 2
- ○ 4 : 1
- ○ 5 : 1

EXPLANATIONS FOR CHARTS AND CRAFTS DRILL

1. C

   There were 200 tons of aluminum and 125 tons of copper purchased in 1985. The sum is 325.

2. B

   There were 175 tons purchased in 1980. 175 tons × \$2,200 per ton = \$385,000.

3. E

   The price per ton of aluminum was \$1,900 in 1975 and \$2,700 in 1985. Therefore, $\frac{difference}{original} \times 100 = \frac{800}{1900} \times 100$ = 42 percent.

4. C

   If Company *X* spent \$3,183,000 on metals and 11 percent of that on iron, then 3,183,000 × .11, or \$350,130, was spent on iron. Divide \$350,130 by 125 tons of iron to find that the price per ton of iron is approximately \$2,800.

5. B

   Statements (A) and (C) cannot be inferred because there is a lack of information, but Statement (B) must be true. Although the ratio of aluminum spending to copper spending is 36 to 24, the ratio of amount of aluminum purchased to amount of copper purchased (295 tons to 175 tons) exceeds the 36 to 24 ratio. This means that copper cost more per ton than did aluminum.

6. B

   The bar graph indicates that Company *Y* had sales of approximately \$235 million in 1988.

7. C

   The bar graph shows that Company *W* and Company *X* have more than 4,000 employees.

8. C

   Set up fractions of revenue over employees: Company $W = \frac{375}{5.1}$, Company $X = \frac{470}{7.9}$, Company $Y = \frac{235}{2.2}$, and Company $Z = \frac{305}{3.8}$. Company *Y* has highest revenue from sales per employee.

9. B

   Company *W* had \$375 million in sales, *X* had \$470 million, *Y* had \$235 million, and *Z* had \$305 million. The average is the total (375 + 470 + 235 + 305 = 1,385) divided by 4, or approximately \$345 million.

10. D

    Stereo receipts were \$72 million, and computer receipts were \$176 million. The percent change is $\frac{difference}{original} \times 100 = \frac{104}{72} \times 100$, which is 144 percent.

11. A

Foodstuffs make up about $51 billion of the $270 billion total Scandinavian exports. $\frac{51}{270} \approx \frac{1}{5} \approx \frac{20}{100}$; therefore, foodstuffs make up approximately 20 percent of total exports.

12. E

Sweden, Denmark, and Finland account for $54 billion in machinery exports out of a total of $62 billion. Therefore, Norway must account for the remaining $8 billion.

13. E

Denmark accounts for approximately 25 percent of the revenue from total Scandinavian exports, but you cannot assume that Denmark accounts for 25 percent of all the different products. There is no way to determine what Denmark produces for export.

14. C

Revenue from all Swedish exports was approximately $132 billion, and 15 percent of 132 is $19.8 billion. Revenue from Norwegian food stuffs is 40 percent less than $19.8 billion, or approximately $12 billion. $12 billion out of a total of $42 billion is approximately 28 percent.

15. E

If $1 buys you 10 pounds of timber, then $103 billion buys you about 1,000 billion pounds of timber. Similarly, if $1 buys you 4 pounds of foodstuffs, then $51 billion buys you about 200 billion pounds of foodstuffs. Therefore, the ratio of pounds exported timber to pounds exported foodstuffs is 1,000 to 200, or 5 to 1.

Chapter 6
Math in the Real World

TRYING TO RELATE

The most common critique against the GRE is that it tests topics that have no direct connection to the topics most graduate school students will use. So the Educational Testing Service, makers of the GRE, decided not to test too much abstract math. Instead, they tried to make questions have actual, real-life connections.

Of course, for people that spend their entire day writing standardized tests, "real-life situations" means something very different than it does for most normal people. For the GRE, this is a real-world problem:

Question 11 of 20

Rhoda put money into her savings account for four consecutive months: March, April, May, and June. Over the last three months, the amount she saved each month was double the amount in the previous month. If the amount she saved in June was \$222.50 more than the amount she saved in March, how much did she save in March?

- ○ \$260.00
- ○ \$130.00
- ○ \$97.50
- ○ \$65.00
- ○ \$32.50

Here's How to Crack It

A typical, real-world situation that we've all been in. You save and save, but then forget how much you have in your account. When you ask your bank how much you have, you are informed that you put \$222.50 more in your account this month than you did three months ago. "But how much do I actually have in my account?" you ask. The bank teller types some numbers into her computer, answers "you put double the amount into your account each month for the past four months. Since that answers your question, thank you for banking with Oblique Bank, and remember to sign up for an Obfuscation Checking Account," and hangs up.

You mean you've never been in that situation? But it's a perfectly *common* situation.

Okay, so it's not a common situation at all. However, it is a common style of GRE question, and one that you may have seen before. Notice all those numbers in the answers, and how the question is asking for a specific amount? You may have recognized the opportunity to PITA. If so, good for you. If not, feel free to look over Chapter 4 to learn more about Plugging In the Answers.

Write down A B C D E on your scratch paper, copy the answers, and label the column "March." Start with (C). If she saved \$97.50 in March, then she saved double that in April (\$195), doubled again in May (\$390), and doubled once more in June (\$780). The amount she saved in June was (\$780 – \$97.50) = \$682.50 more than what she saved in March, which is way more than \$222.50. Because (C) is too big, cross off (C), (B), and (A).

Let's try (D). If she saved \$65 in March, she saved \$130 in April, \$260 in May, and \$520 in June. She therefore spent (\$520 – \$65) = \$455 dollars more in June than in March, which means (D) is too large as well. Cross off answer choice (D), and pick (E), the only answer left.

Fictional World Examples

So the questions on the GRE won't actually apply to the real world. However, they'll use certain real-world topics frequently. You should know exactly what to do when any of these topics come up, so you don't get lost in all the extra word problem garbage.

As always, focus on learning what you need to recognize in a problem to know what to start writing on your scratch paper.

THE THREE M'S

When it comes to crunching numbers, there are three statistical terms that every GRE student should recognize and distinguish from each other.

- The **mean**, or "arithmetic mean," is just another word for the average.
- The **median** is the middle number in a set.
- The **mode** is the term that occurs *most* frequently in a set of numbers.

To remember these last two terms, you can think that (1) the *median* of a highway is in the *middle* of the highway, and (2) if you say the word *most* as if you have a terrible head cold, it comes out sounding like *mode.*

One final term you should know is **range**; the range of a set of numbers is the difference between the greatest and smallest numbers in the set.

The Mean (Average)

You calculate the average value of a set of numbers by finding their total value and dividing by the quantity of numbers in that set. To find the average of 12,

29, 32, 8, and 19, for example, add them all up (12 + 29 + 32 + 8 + 19 = 100) and divide by the number of terms (five). The answer is $\frac{100}{5}$, or 20.

Question 9 of 20

A basketball player scores 12 points during her first game, 29 during her second, 32 in her third, 8 in her fourth, and 19 in her fifth game. What was her average score per game during that week?

- 14
- 17
- 20
- 25
- 33

Here's How to Crack It

From our previous work, we know that the answer is (C). However, it's worth pointing out that you can eliminate a few answer choices right away. In particular, choice (E) stands out, because the average of a set of numbers cannot be greater than the greatest number in the set. To solve the problem, you can use the Average Pie.

The Average Pie

All average problems involve three quantities—the *Total* value, the *Number* of elements, and the *Average* value of those elements. You can relate them in a diagram we call the Average Pie, which looks like this:

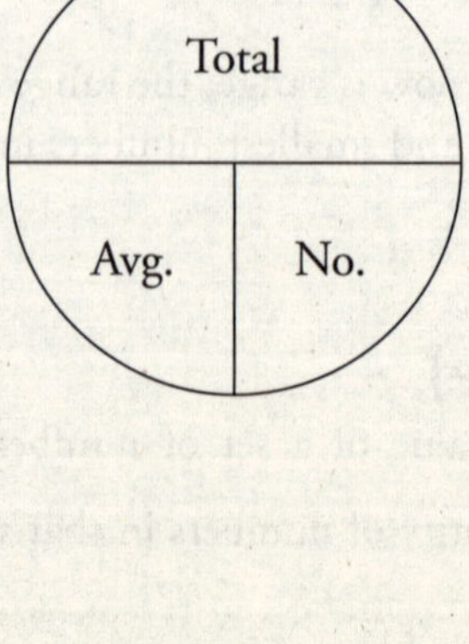

The Average Pie helps you visualize the relationship between the three numbers. It also helps you organize your thoughts by giving you three discrete compartments in which to put your information.

In order to solve the previous problem, you would add up the elements to get 100, recognize that there were five numbers, and place that information in the Average Pie like this:

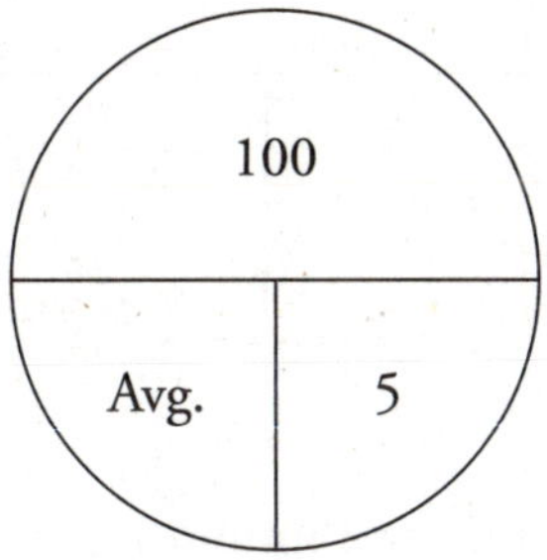

When you divide 100 by 5, you see that the answer is 20.

Try another example:

Question 14 of 20

During the six weeks between Thanksgiving and New Year's Eve, an average of 12.6 million people rode mass transit in City X each week.

Quantity A	**Quantity B**
The number of people who rode mass transit in City X between Thanksgiving and New Year's Eve	75 million

- ○ Quantity A is greater.
- ○ Quantity B is greater.
- ○ The two quantities are equal.
- ○ The relationship cannot be determined from the information given.

Trigger: The word "average."

Response: Draw an average pie for every time the word average appears in the question.

Here's How to Crack It

Here you have two of the three elements that go in the Average Pie. You know the Average and the Number, and you're looking for the Total, so set up your Average Pie like this:

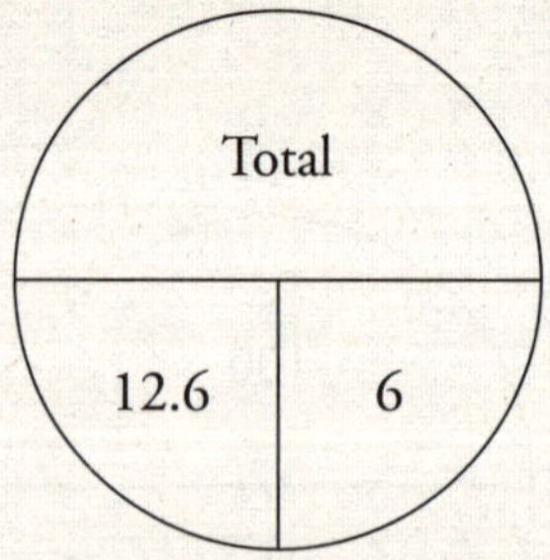

To find the Total, multiply the two bottom numbers: $6 \times 12.6 = 75.6$ million, which is slightly larger than the 75 million in Column B. The answer is (A).

Averages Quick Quiz

Question 1 of 4

Day	# of Eggs
Sunday	43
Monday	56
Tuesday	39
Wednesday	31
Thursday	46
Friday	49
Saturday	37

Farmer Brown took note of the quantity of eggs produced by the hens on his farm and compiled the above chart. What was the average number of eggs that his hens produced each day?

- ○ 43
- ○ 44
- ○ 45
- ○ 46
- ○ 47

Question 2 of 4

On average, each laying hen lays 255 eggs per year. If Farmer Jones has 21 hens, how many eggs can she expect to get over the course of a year?

[ ]

Click on the answer box, then type in a number. Backspace to erase.

Question 3 of 4

During a particularly productive month, Farmer Green's hens laid a total of 572 eggs. If each hen laid an average (arithmetic mean) of 22 eggs that month, how many hens did Farmer Green have that month?

- ○ 18
- ○ 22
- ○ 26
- ○ 32
- ○ 45

Question 4 of 4

Over a certain month, the average (arithmetic mean) of eggs laid among 12 hens was 9. If Farmer Dobbs removes the eggs laid by the top three producers, the average of the remaining hens is 7.

Quantity A	**Quantity B**
The average number of eggs laid by the top three producers	11

- ○ Quantity A is greater.
- ○ Quantity B is greater.
- ○ The two quantities are equal.
- ○ The relationship cannot be determined from the information given.

Explanations for Averages Quick Quiz

1. During the seven days, the hens produced a total of 301 eggs. The Average Pie, therefore, looks like this:

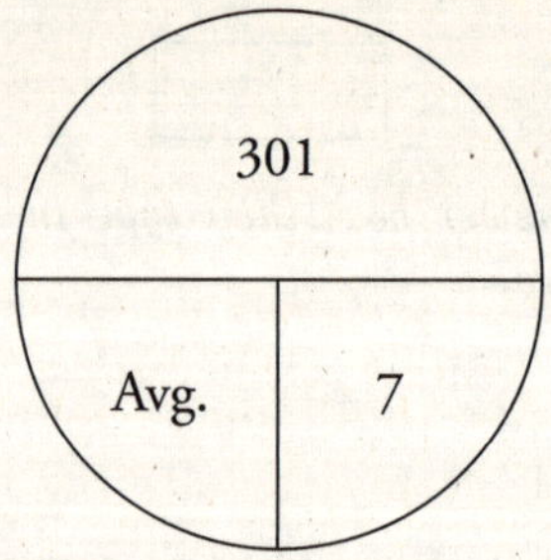

The average value equals 301 ÷ 7, or 43. The answer is (A).

2. You know the number and the average, so set up your Average Pie like this:

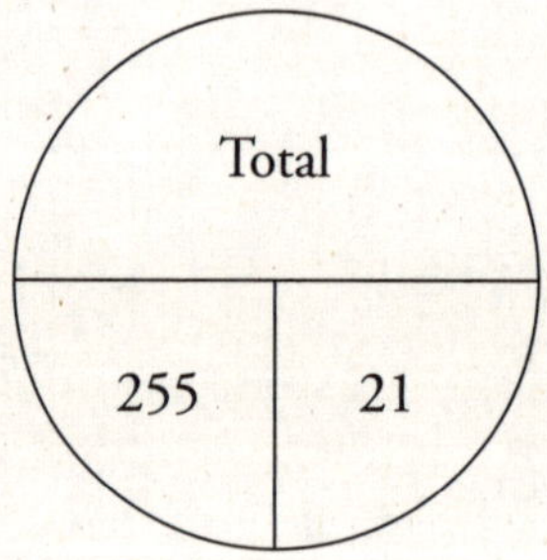

Farmer Jones can expect to get 21 × 255, or 5,355 eggs over the course of a year.

3. This time, you know the total and the average, so the Average Pie looks like this:

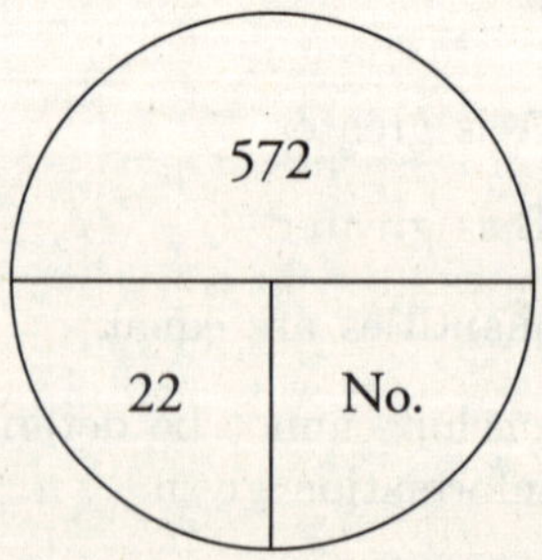

The number of hens equals 572 ÷ 22, or 26 hens. The answer is (C).

4. If 12 hens lay an average of 9 eggs apiece, then the total number of eggs is 12 × 9, or 108. Once the top three egg-layers are disqualified, there are 9 hens left that lay an average of 7 eggs. They account for 7 × 9, or 63 of the eggs. The top three layers must therefore account for 108 – 63, or 45 eggs. The average value among the top three is 45 ÷ 3, or 15. The Average Pies are laid out here:

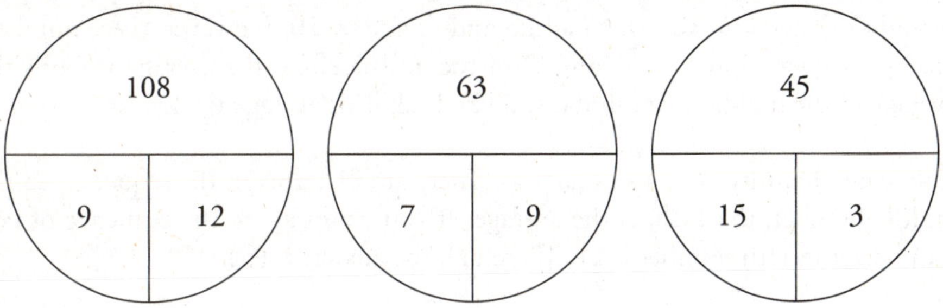

Because 15 is greater than 11, the answer is (A).

Sequences: A Helpful Hint

The GRE likes to make certain average questions seem more difficult and time-consuming than they are by having them involve huge sequences of numbers. The good news is that if the elements in a set do follow a sequential pattern, there's a lot less work involved than you might think.

The average of any sequence of elements is either

- the middle number (if the number of elements is odd); or
- the average of the middle two numbers (if the number of elements is even).

Question 8 of 20

Quantity A	**Quantity B**
The average value of the first 20 even numbers	The average value of the first 21 odd numbers

○ Quantity A is greater.

○ Quantity B is greater.

○ The two quantities are equal.

○ The relationship cannot be determined from the information given.

Here's How to Crack It

The decoy answer is (B), because it looks like 21 numbers would lead to a greater answer than 20 numbers. Keep in mind that because there are no variables, you can eliminate (D).

The first 20 even numbers are 2, 4, 6, 8, 10, 12, 14, 16, 18, 20, 22, 24, 26, 28, 30, 32, 34, 36, 38, and 40. Once you listed them all out, you might panic at the thought of having to add them all up and divide by 20. However, these numbers are in a sequence, and there are 20 of them. Therefore, the average value is the average of the middle two numbers, 20 and 22. This average is 21.

Calculate Quantity B in a similar way. There are 21 terms in the sequence, so the middle number, the 11th, is the average. If you count along the sequence of odd numbers, the 11th number is 21. Therefore, the answer is (C).

The Median

As you know, the median of a set of elements is the middle value when the numbers are placed in order of increasing size. One of the most common places to find median values is in a grad-school brochure, which often displays its "median" GRE score. Because the average of so many numbers will most likely end in a decimal, the median is preferable because it's much more likely to be an integer.

Trigger: The word "median" appears in the problem.

Response: Put the list of numbers in order and find the middle number.

Once again, the number of elements in the set is important. Once you've ordered them from least to greatest, the median will either be the middle value (if the number of elements is odd) of the average of the middle two values (if the number of elements in the set is even).

Question 12 of 20

Set D = {the first 13 positive integers}

Set F = {the prime elements in Set D}

Quantity A	**Quantity B**
The median value of Set D	The median value in Set F

- ○ Quantity A is greater.
- ○ Quantity B is greater.
- ○ The two quantities are equal.
- ○ The relationship cannot be determined from the information given.

Here's How to Crack It

Set D contains 15 elements (numbers 1 through 15), so the median is the middle number, or 7. The prime elements in Set D are 2, 3, 5, 7, 11, and 13, so the median is the average of the middle two numbers, 5 and 7. Their average is 6, so the answer is (A).

Three M's Quick Quiz

Set G comprises the numbers {4, 8, 5, 9, 8, 3, 8, 5, 4}

Question 1 of 5

What is the average (arithmetic mean) value of Set G ?

[ ]

Click on the answer box, then type in a number. Backspace to erase.

Question 2 of 5

Quantity A	**Quantity B**
The median value of set G	5

- ○ Quantity A is greater.
- ○ Quantity B is greater.
- ○ The two quantities are equal.
- ○ The relationship cannot be determined from the information given.

Question 3 of 5

Quantity A	**Quantity B**
The mode value of set G	The median value of set G

- ○ Quantity A is greater.
- ○ Quantity B is greater.
- ○ The two quantities are equal.
- ○ The relationship cannot be determined from the information given.

Question 4 of 5

What is the range of Set *G*?

[ ]

Question 5 of 5

If another element is added to Set *G* and the median value doesn't change, which of the following could be that element?

*Indicate **all** possible values.*

- ☐ 1
- ☐ 3
- ☐ 4
- ☐ 5
- ☐ 6
- ☐ 7
- ☐ 9

Explanations for Three M's Quick Quiz

1. The sum of all of the elements in Set *G* is $4 + 8 + 5 + 9 + 8 + 3 + 8 + 5 + 4$, or 54. There are nine elements, so the average value is $54 \div 9$, or 6.
2. When the elements are arranged in order, Set *G* looks like this: {3, 4, 4, 5, 5, 8, 8, 8, 9}. The middle value is 5, so the answer is (C).
3. The element that occurs most often is 8, which is greater than the median (5). The answer is (A).
4. The least value in Set *G* is 3, and the greatest value is 9. Therefore, the range of the set is $9 - 3$, or 6.
5. Any new value that is added will make the number of elements an even 10, so the new median will be the average value of the fifth and sixth elements. Any number that is less than or equal to 5 will make fifth and sixth elements both 5, effectively keeping the median at 5. Therefore, any of the first four values (1, 3, 4, and 5) will work.

RATIOS AND PROPORTIONS

Ratios are a lot like fractions and decimals, with one important difference: Fractions and decimals compare parts to the whole, while ratios are more concerned with comparing two or more parts that together don't necessarily represent the whole. Most of the time, ratios are denoted with a colon, as in "the ratio of boys to girls in the classroom was 4 : 3." This means that for every four boys in the room, there were three girls. The actual number of boys is therefore a multiple of 4, the number of girls is a multiple of 3, and the number of children is a multiple of 7.

Most of us have been trained to use algebra when solving ratio questions, but the Ratio Box lets you throw algebra out the window.

The Ratio Box

Rather than deal with variables when you encounter a ratio problem, you can use the Ratio Box to organize your data in nice little columns:

Trigger: The word "ratio" appears in the problem.

Response: Draw a Ratio Box on your scratch paper.

	PART	PART	WHOLE
Ratio			
Multiplier			
Total			

The Ratio Box is a great tool because it lets you compare the parts within the whole at a glance and it clearly relates the ratio (along the top row) to the actual number of elements you have (the bottom row). Here's how to use it:

Question 7 of 20

In a certain used-car showroom, the ratio of hardtop cars to convertibles is 7 : 2. If there are 16 convertibles, how many cars are in the showroom?

○ 2

○ 7

○ 9

○ 56

○ 72

Here's How to Crack It

Set up your Ratio Box, label your Parts columns, and enter all the information you know:

	Hardtops	Convertibles	Whole
Ratio	7	2	9
Multiplier			
Total		16	

Notice that the first thing to do with the numbers in the ratio row is to put their sum in the far-right column. Now, from the top row, we know that for every 9 cars, 7 of them are hardtops and 2 are convertibles.

The next step is to make the connection between the ratio of the cars and the actual number of cars, which are separated by a multiplier. The link is in the convertible column; there are 16 actual convertibles, and the ratio value is 2. Therefore, the multiplier for the whole box is 16 ÷ 2, or 8. Enter 8 across the entire multiplier row, like this:

	Hardtops	Convertibles	Whole
Ratio	7	2	9
Multiplier	8	8	8
Total	56	16	72

Finish the box by multiplying down each of the other columns.

We now know that there are 56 hardtops and 16 convertibles, for a total of 72 vehicles. (Note that if you've done your math correctly, the actual amounts in the Parts columns should add up to the actual amount in the Whole column.) The answer to our question is 72, and the credited response is (E).

It's Expandable!

Not all ratio questions have just two parts, and you can expand your Ratio Box to include as many parts as necessary. Here's how that works:

Question 9 of 20

A contractor mixes cement, water, sand, and gravel in a ratio of 1 : 3 : 4 : 2 to make concrete. In order to make 28 pounds of the mixture, how much sand will he need?

- ○ 2.8
- ○ 5.6
- ○ 8.4
- ○ 11.2
- ○ 13.6

Here's How to Crack It

The formula has four parts, so create a column for each one, like this:

	C	W	S	G	Total
Ratio	1	3	4	2	10
Multiplier			2.8		2.8
Total			11.2		28

The total is 1 + 3 + 4 + 2, or 10 parts. If the actual amount is 28 pounds, then the multiplier is 28 ÷ 10, or 2.8. Don't bother filling in every column; all you need is the amount of sand. Place the 2.8 in the sand column, multiply by 4, and you get 11.2 pounds. The answer is (D).

Ratios Quick Quiz

Question 1 of 4

The ratio of boys to girls in a hospital maternity ward is 3 : 4. Which of the following CANNOT be the actual number of babies in the ward?

- ○ 28
- ○ 35
- ○ 38
- ○ 42
- ○ 63

Question 2 of 4

If there are only tulips and daffodils in Uncle Henry's garden and the ratio of tulips to daffodils is 3 : 5, what percent of the flowers in the garden are daffodils?

- ○ 37.5%
- ○ 40%
- ○ 60%
- ○ 62.5%
- ○ 133.3%

Question 3 of 4

In a certain restaurant, the ratio of line cooks to waiters is 2 : 3 and the ratio of waiters to busboys is 4 : 3.

Quantity A	**Quantity B**
The number of line cooks in the restaurant	The number of busboys in the restaurant

- ○ Quantity A is greater.
- ○ Quantity B is greater.
- ○ The two quantities are equal.
- ○ The relationship cannot be determined from the information given.

Question 4 of 4

If there are 120 workers in an office, each of the following could be the ratio of men to women EXCEPT

- 2 : 1
- 3 : 1
- 4 : 1
- 6 : 1
- 7 : 1

Explanations for Ratios Quick Quiz

1. When you fill in the Ratio Box with the given ratio, the sum of the two parts is 7.

	Boys	Girls	Whole
Ratio	3	4	7
Multiplier			
Total			38

Therefore, the number of babies in the maternity ward must be a multiple of 7. Each of the answer choices except 38 is a multiple of 7. The answer is (C).

2. If the ratio of tulips to daffodils is 3 : 5, your Ratio Box looks like this:

	Tulips	Daffodils	Whole
Ratio	3	5	8
Multiplier			
Total			

Because the total is 8, the fractional amount of daffodils is 5 out of 8, or $\frac{5}{8}$. Because $\frac{5}{8}$ is equivalent to 62.5 percent (use your calculator if you haven't already memorized this), the answer is (D).

3. Line up your ratios like this, so that you can compare all three parts at once.

line cooks		waiters		busboys
2	:	3		
		4	:	3
8	:	12	:	9

There are two separate ratios listed here, and the only way to compare line cooks and busboys is to relate them to a common number of waiters. If you multiply the two numbers you have for waiters, you get 12—a lowest common denominator. From here, you can convert the ratio of line cooks to waiters from 2 : 3 to 8 : 12, and the ratio of waiters to busboys from 4 : 3 to 12 : 9. This makes the ratio of line cooks to busboys 8 : 9. There are more busboys than line cooks, so the answer is (B).

4. When you fill in the Ratio Box, the value in the Whole column of the ratio row must be a factor of 120. Because 6 + 1 = 7, and 7 is not a factor of 120, the answer is (D).

Proportions

Proportions are related to ratios, because a ratio between two elements is proportional to the actual values. If you know the ratio of two quantities and you have to extrapolate that ratio onto some actual values, you'll probably end up writing a proportion.

The Setup

Proportions are comparisons of two things in a fixed ratio. There are two very important considerations to keep in mind when you set proportions up.

- **Make sure your elements are consistent:** If you're comparing miles per hour and you decide that miles are in the numerator, make sure they're always in the numerator.
- **Make sure your units are consistent:** If you're comparing distances, and one distance is given in feet while the other is in inches, convert one of those distances so that the units are the same.

Once your proportion is ready, you can cross-multiply and solve for the missing value.

Question 5 of 20

If a train maintains a constant speed of 80 miles per hour, how far will the train travel during the first 24 minutes?

Here's How to Crack It

The first ratio compares miles and hours, so your first instinct might be to set up a proportion like this:

$$\frac{miles}{hours} \quad \frac{80}{1} = \frac{x}{24}$$

The elements are aligned, but the units aren't consistent because the question mentions 24 *minutes*, not hours. To solve this, convert 1 hour to 60 minutes, and you'll be ready to cross-multiply:

$$\frac{miles}{minutes} \quad \frac{80}{60} = \frac{x}{24}$$

$$60x = 80 \times 24$$

$$60x = 1{,}920$$

$$x = 32$$

Proportions Quick Quiz

Question 1 of 4

A muffin recipe that calls for 2 cups of flour yields 25 muffins. If Yvonne needs to make 40 muffins, how much flour should she use?

- ○ $1\frac{1}{4}$
- ○ $2\frac{1}{3}$
- ○ $2\frac{4}{5}$
- ○ $3\frac{1}{5}$
- ○ $5\frac{1}{3}$

Question 2 of 4

A greenskeeper wants to fertilize the eighteenth hole of a golf course, which has an area of 12,000 square feet. If the fertilizer he wants to use suggests using one bag for 10 square yards, how many full bags will he have to buy?

- ○ 48
- ○ 133
- ○ 134
- ○ 266
- ○ 1,200

Question 3 of 4

If 75 percent of a certain number is 1,200, then 10 percent of that number is

- ○ 90
- ○ 120
- ○ 160
- ○ 240
- ○ 480

Question 4 of 4

Arturo can save $1,490 in 8 months.

Quantity A	Quantity B
The number of months, at the same rate, it would take Arturo to save 3.5 times that amount	28

- ○ Quantity A is greater.
- ○ Quantity B is greater.
- ○ The two quantities are equal.
- ○ The relationship cannot be determined from the information given.

Explanations for Proportions Quick Quiz

1. Set up the proportion that compares cups of flour to muffins.

$$\frac{2}{25} = \frac{x}{40}$$

When you cross-multiply, your equation becomes $25x = 80$, and $x = 3.2$. Because this is equivalent to $3\frac{1}{5}$, the answer is (D). Note: Just because you can use your calculator doesn't mean you should forget all about fractions, which can still appear in answer choices.

2. First, an important conversion: Because one yard equals three feet, one square yard is the same as *nine* square feet. Therefore, if each bag fertilizes 10 square yards, it fertilizes 90 square feet. Set up your proportion, comparing bags to square feet of coverage:

$$\frac{1}{90} = \frac{x}{12{,}000}$$

When you cross-multiply, $90x = 12{,}000$ and $x = 133.3$. Because he has to buy complete bags, he must buy 134 for complete coverage. The answer is (C).

3. You can find the value of the number, but you really don't need to. Just set up the proportion that compares percentages to numbers.

$$\frac{75}{1,200} = \frac{10}{x}$$

From this proportion, you can cross-multiply to find that $75x$ = 12,000, and x = 160. The answer is (C).

4. Even though the quantities are proportional, there's no need to set up a proportion. If Arturo wants to save 3.5 times the amount, he'll need 3.5 times more time. When you multiply 8 by 3.5, you get 28 months. The two quantities are equal, and the answer is (C).

Naturally, statistics can get a little more complicated, but we'll wait until Chapter 10 to delve into the world of combinations, probability, and standard deviation. In the meantime, here are some more practice questions about ratios, proportions, and real-life math situations.

Math in the Real World Drill

Question 1 of 16

A machine punches x plates per hour for 4 hours and then y plates per hour for 2 hours. What is the average number of plates punched per hour by the machine during the 6-hour period?

- ○ $2x + 4y$
- ○ $\frac{2x + 4y}{6}$
- ○ $\frac{3xy}{2}$
- ○ $\frac{4x + 2y}{6}$
- ○ $4x + 2y$

Question 2 of 16

Twenty bottles contain a total of 8 liters of apple juice. If each bottle contains the same amount of apple juice, how much juice, in liters, is in each bottle?

[ ] liters

Click on the answer box, then type in a number. Backspace to erase.

Question 3 of 16

If 12 melons cost a total of \$9.60, then at the same rate per melon, what is the cost of 9 melons?

- ○ \$7.00
- ○ \$7.20
- ○ \$8.00
- ○ \$8.45
- ○ \$8.65

Question 4 of 16

If $2a = 3b = 4c = 72$, then the average (arithmetic mean) of a, b, and c is

- ○ 39
- ○ 26
- ○ 24
- ○ 18
- ○ 9

Question 5 of 16

Jenny notices a consistent, increasing pattern in the number of geese she sees flying over her house each day. If she sees 5 geese on Monday, 8 geese on Tuesday, 11 geese on Wednesday, and 14 geese on Thursday, and the next week starts on Sunday, which of the following days of the next week will she see a prime number of geese if the pattern continues?

*Indicate **all** possible values.*

- ☐ Sunday
- ☐ Monday
- ☐ Tuesday
- ☐ Thursday
- ☐ Friday
- ☐ Saturday

Click on your choice(s).

Question 6 of 16

Set B contains only positive, even integers. Which of the following could be the median of set B ?

*Indicate **all** such expressions.*

☐ –2

☐ 0

☐ 1

☐ 3

☐ 3.5

☐ 4

Click on your choice(s).

Question 7 of 16

A car manufacturer has 2,992 forklifts, which is approximately one forklift for every 48.9 employees. Which of the following is the closest approximation, in thousands, of the number of employees employed by the manufacturer?

○ 60

○ 100

○ 150

○ 175

○ 300

Question 8 of 16

A recipe for 4 loaves of bread requires $\frac{3}{4}$ cups of sugar. If Chris wants to make 2 loaves of bread, which of the following calculations yields the amount of sugar he needs?

*Indicate **all** that apply.*

☐ $\frac{3}{4} \times \frac{1}{2}$

☐ $\frac{3}{4} \times 2$

☐ $\frac{3}{4} \div \frac{1}{2}$

☐ $\frac{3}{4} \div 2$

Click on your choice(s).

Question 9 of 16

V is a sequence of numbers in which every term after the first two is the average of the two previous terms. If the first term x is 16 more than the second term y, then which of the following represents the fifth term of the sequence, in terms y ?

○ $y - 2$

○ y

○ $y + 2$

○ $y + 4$

○ $y + 6$

Question 10 of 16

Set A contains only even integers. Which of the following CANNOT be the median of set A ?

○ –2

○ –1

○ 0

○ 0.5

○ 1

Question 11 of 16

A violinist needs 2 hours to tune a violin made in the twentieth century. To tune violins made before the twentieth century, the violinist needs twice as long, and for violins made after the twentieth century, she needs half as long. Which of the following groups of violins could she tune in 6 hours?

*Indicate **all** that apply.*

☐ Two violins made before the twentieth century

☐ Three violins made during the twentieth century

☐ One violin made during the twentieth century, one made before, and two made after

☐ Two violins made after the twentieth century, and one made before

☐ Two violins made after the twentieth century, and two made during the twentieth century

Click on your choice(s).

Question 12 of 16

$$2g = 6h$$

Quantity A	**Quantity B**
The ratio of g to h	$\frac{1}{3}$

○ Quantity A is greater.

○ Quantity B is greater.

○ The two quantities are equal.

○ The relationship cannot be determined from the information given.

Question 13 of 16

A class of 5 students took an exam and scored an average of x. When a sixth student took the test and scored a 93, the class average was 88. What is the value of x ?

○ 84

○ 85

○ 86

○ 87

○ 89

Question 14 of 16

Let H be a sequence, h_1, h_2, h_3 ... h_4, such that each term after the first is two less than one-third of the previous term. If the fourth term in the sequence is 0, which of the following is the sum of the first and fifth term of the sequence?

○ –2

○ $25\frac{1}{3}$

○ 26

○ 76

○ 78

Question 15 of 16

A bag of jellybeans has red and yellow jellybeans in a ratio of $c : d$. If there are r red jellybeans in the bag, which of the following represents the number of yellow jellybeans?

- ○ $\frac{cd}{dr}$
- ○ $\frac{dr}{c}$
- ○ $\frac{cd}{r}$
- ○ $d(c + r)$
- ○ $d(r - d)$

Question 16 of 16

Miguel's bowling team bowled a practice round in preparation for their upcoming league game. The team's average score for the practice round was 180. Miguel scored 190, Janice scored 200, and Thad scored 210. If no team member scored less than 165, and none of the remaining team members scored above 170, what is one possible value for the number of members on Miguel's team? (Note: Bowling scores are always positive integers.)

[ ]

EXPLANATIONS FOR MATH IN THE REAL WORLD DRILL

1. D

   Plug In, and let $x = 6$ and $y = 3$. If the machine punches 6 plates per hour for 4 hours, you know that it punches 24 plates in that time. If the machine punches 3 plates per hour for 2 hours, you know that it punches 6 plates in that time. In 6 hours the machine punches a total of 30 plates. If you divide 30 by 6 you'll get 5 plates per hour, your target answer. The only answer that matches the target is answer choice (D): $\frac{(4\times 6)+(2\times 3)}{6} = \frac{24+6}{6} = 5$.

2. 0.4

   Twenty bottles hold a total of 8 liters of juice. Each bottle contains the same amount, so you need to divide the amount of juice by the number of bottles. $8 \div 20 = 0.4$.

3. B

   Divide the total cost of \$9.60 by 12 melons to get the per-melon cost of \$0.80. Now calculate the cost of 9 melons at the same rate: 9 × \$0.80 = \$7.20.

4. B

   Break the expression into three equations. $2a = 72$, so $a = 36$. $3b = 72$, so $b = 24$. $4c = 72$, so $c = 18$. The average of 36, 24, and 18 is $\frac{36+24+18}{3} = 26$. You could estimate that choice (E) is incorrect because the value of each of the three variables is greater than 9, and the average will be greater than 9.

5. A, C, and F

   Once you deduce the pattern, you'll be able to predict the number of goose sightings for each day of next week and select the prime numbers. The number of goose sightings increases by 3 each day: If she saw 14 on Thursday, she'll see 17 on Friday, 20 on Saturday, 23 on Sunday, 26 on Monday, 29 on Tuesday, 32 on Wednesday, 35 on Thursday, 38 on Friday, and 41 on Saturday. Sunday, Tuesday, and Saturday have a prime number of goose sightings, so choices (A), (C), and (F) are correct.

6. D and F

   If the set contains only positive integers, then there's no way numbers 0 or smaller can be the median, so eliminate choices (A) and (B). For 1 to be the median, there has to be a number in the set smaller than 1, so eliminate choice (C). Don't eliminate choice (D) just because it's odd—remember that in a set with no unique middle term, the median will be the average of the two middle terms; the set could be {2, 4}, for instance, which would have a median of 3. The same set could be extended to {2, 4, 6}, which would have a median of 4, or choice (F). Since 3.5 itself cannot be in the set, the only remaining question is whether 3.5 can be the average of two even numbers: it can't, because 3.5 × 2 = 7, and two even numbers can't add up to 7. Eliminate choice (E) and select choices (D) and (F).

7. C

   Estimate. 2,992 is almost 3,000, and 48.9 is almost 50. The number of employees is approximately 3,000 × 50, or 150,000. Only choice (C) is close.

8. A and D

Because Chris is only making 2 loaves, he needs half as much sugar as he does for 4 loaves. You can multiply $\frac{3}{4}$ by $\frac{1}{2}$ to find how much sugar he needs, or you can divide the quantity by 2.

9. E

See variables in the answer choices, plug in. Start by choosing your *y*. *x* is 16 more than *y*, so let *y* be some small positive integer like 2. That makes $x = 18$. Average the two together, and the third term in the sequence (remember, *x* and *y* are the first *two* terms of the sequence) is 10. Take your second and third terms, average them together, and you should get 6 ($(10 + 2) \div 2 = 6$) as your fourth term. Do it one more time to get 4 ($(6 + 10) \div 2 = 8$) as the fifth term. Plug in to your answer choices, and eliminate all but choice (E).

10. D

Come up with possible sets that could have the medians of the answer choices. Choices (A) and (C) could work because –2 and 0 are even numbers and you could easily have a set with those choices as the center number. Eliminate these choices. Both –1 and 1 are a bit more difficult to eliminate, but keep in mind that if your set has two middle terms, rather than just one, the median of the set is the average of those two middle terms. So, for a median of –1, the set could be –2 and 0; for a median of 1, the set could be 0 and 2. Eliminate choices (B) and (E). Select your only remaining choice (D).

11. B, D, and E

First, figure out how long it will take the violinist to tune each type of instrument. Violins made before the twentieth century take twice as long to tune, so they take 4 hours. Violins made after the twentieth century take half as long to tune, so they take 1 hour. Add up the total time needed to tune the violins in each answer choice. The total time for answer choice (A) is $4 + 4 = 8$ hours, so choice (A) can be eliminated. For answer choice (B), the total time is $3 \times 2 = 6$ hours. Choice (B) is correct. For answer choice (C), $2 + 4 + 1 + 1 = 8$ hours. Choice (C) can be eliminated. For answer choice (D), $1 + 1 + 4 = 6$ hours. Choice (D) is correct. For answer choice (E) $1 + 1 + 2 + 2 = 6$ hours. Choice (E) is correct.

12. A

Since there are variables involved, this is an opportunity to plug in. First, simplify the equation by dividing both sides by 2 to get $g = 3h$. Plug in a number for *h* and solve for *g* to see the ratio. For example, if you plug in 1 for *h*, then $g = 3$. Quantity A asks for the ratio of *g* to *h*, which is now 3 to 1. Quantity A is greater.

13. D

Using the average wheel, the sum of the first 5 students' scores = $5x$. The sum of all 6 students' scores then is $5x + 93$, the old sum plus the new score. The average of all 6 scores is $\frac{5x+93}{6} = 88$. Multiply both sides by 6 and solve to find $x = 87$. Choice (D) is the best answer.

14. **D**

With sequence questions, PITA typically works well. This problem makes that a little more difficult by asking you for a sum of two terms, rather than the term itself (i.e. you would still have to guess about the terms, even if you attempted to begin at a certain answer choice and work from there). You are given the fourth term, so try to move back and forth from there. Each following term is one-third of the previous term minus 2. So, $\frac{1}{3}$ of 0 equals 0, and 0 – 2 equals –2, your fifth term. To find the previous terms, simply invert the sequence: add 2, then multiply by 3. The third term is $(0 + 2) \times 3 = 6$, the second term is $(6 + 2) \times 3 = 24$, and the first term must be $(24 + 2) \times 3 = 78$. Add together your first and fifth terms: $78 + -2 = 76$, which is choice (D).

15. **B**

Use the ratio box and lug in some values. If $c = 2$, $d = 3$, and $r = 20$, then there will be 30 yellow jellybeans. Circle 30 as your target and check all five choices. Choices (A) and (C) are way too small, but choice (B) works. Choices (D) and (E) do not match, leaving choice (B) as your correct answer.

16. **7, 8,** or **9**

Use the Average Pie. Label the number of additional team members t. Assume that all of these t people scored 165. These t people therefore scored a total of $165t$ points. Miguel, Janice, and Thad scored a total of 600 points. Therefore the total number of points scored by the team is $165t + 600$. Put this in the top segment of the pie. The total number of people on the team is $t + 3$, which should go in the bottom left segment. Put the average, 180, in the bottom right segment. At this point you can see that $\frac{165t + 600}{t + 3} = 180$, or total divided by number of things equals average. Solve algebraically. First, $165t + 600 = 180t + 540$. Subtract 540 from both sides to find that $165t + 60 = 180t$. Subtract $165t$ from both sides to find that $60 = 15t$. Therefore, $t = 4$. Remember that t is the number of *additional* people on the team, and the question asks for total people, therefore if $t = 4$, there are 7 people on the team. The problem also gives you an integer for t if you use 170 ($t = 6$, and the total number of team members is 9) and 168 ($t = 5$).

Chapter 7
Geometry

ANGLING FOR A BETTER GRADE?

Now that you've reached this chapter, you may be having a flashback to your freshman year in high school when you first came in contact with theorems, postulates, and definitions, all woven together to form the geometric proof. Well, relax. GRE geometry questions have little to do with deductive reasoning. You're much more likely to be tested on the basic formulas involving area, perimeter, volume, and angle measure. As you work through the GRE math, you'll find that there is a basic battery of terms and formulas that you should know for the geometry questions that do come your way. Before we get to those, let's look at some techniques.

- **Plug In.** If a problem tells you that a rectangle is x inches long and y inches wide, plug in some real numbers to help the question take on a more tangible quality. (And if there is more than one variable, remember to plug in a different number for each one.)
- **Use Ballparking.** If a diagram is drawn to scale, you can sometimes estimate the right answer and eliminate all the answer choices that don't come close.
- **Re-draw to Scale.** If a diagram is not drawn to scale—and you'll know because you'll see the words "Note: Figure not drawn to scale" right below the picture—re-draw it to make it look like it's supposed to look like. Drawings like this are meant to confuse you by suggesting that the figure looks how it's represented in the problem. Re-drawing the figure to scale helps you avoid falling into that trap. So remember (and this is very important): *The diagram isn't drawn to scale unless the problem specifically states that it is!*

BASIC HINTS FOR GEOMETRY QUESTIONS

Geometry is a special science all its own, but that doesn't mean it marches to the beat of an entirely different drummer. Many of the techniques you've learned for other problems will work here as well.

So let's start with our pal Euclid and his three primary building blocks of measured space: points, lines, and planes.

LINES AND ANGLES

Two points determine a line, and two intersecting lines form an angle measured in degrees. There are 360° in a complete circle, so halfway around the circle forms a straight angle, which measures 180°, and half of that is a right angle, which measures 90°.

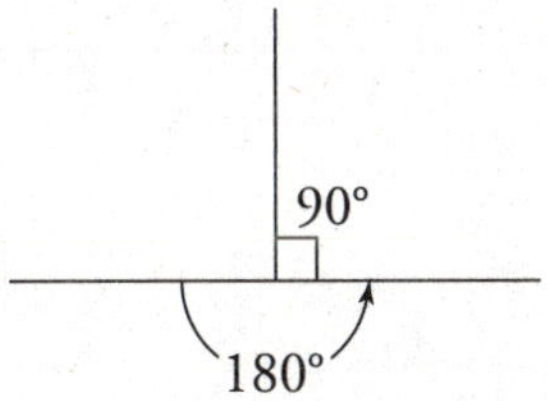

Two lines that intersect in a right angle are perpendicular, and perpendicularity is denoted by the symbol "⊥." Two lines that lie in the same plane and never intersect are parallel, which is denoted by "||." Take a look at two parallel lines below:

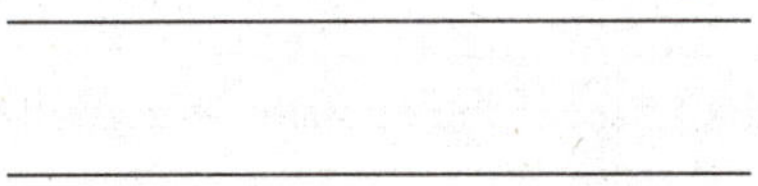

If one line intersects two parallel lines, it is called a transversal. It may look as though this transversal creates eight angles with the two lines, but there are actually only two types: big ones and small ones. All the big ones have the same measure, and all the small ones have the same measure. The sum of the measures of one big angle and one small angle is always 180°.

Trigger: Two parallel lines cut by a transversal.

Response: Label all acute (small) angles as equal, and all obtuse (large) angles as equal.

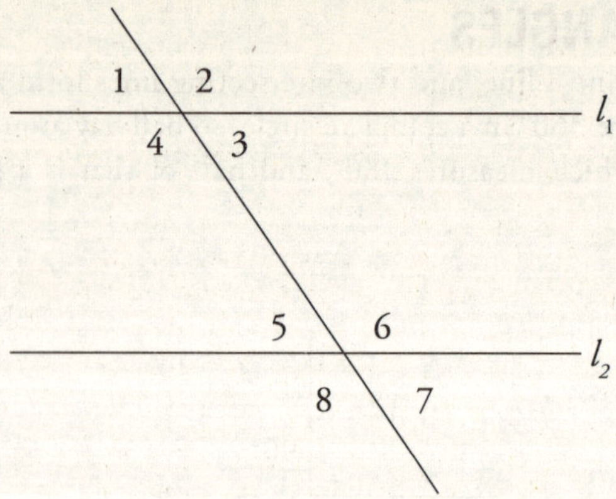

$l_1 \| l_2$

Notice that for the figure above, the acute (small) angles labeled 1, 3, 5, and 7 are all the same, because l_1 is parallel with l_2. We also know that the angles labeled 2, 4, 6, and 8 are all the same for the same reason. Whenever the GRE states that two lines are parallel, look to see if the question is actually testing this concept.

TRIANGLES

Three points determine a triangle, and all triangles have three sides and three angles (but you already knew that). The sum of the measures of the angles inside a triangle is 180°. The sides and angles are related (you don't have to know why, though, because there isn't any trigonometry on the test). Just remember that the longest side is always opposite the largest angle, and the shortest side is always opposite the smallest angle.

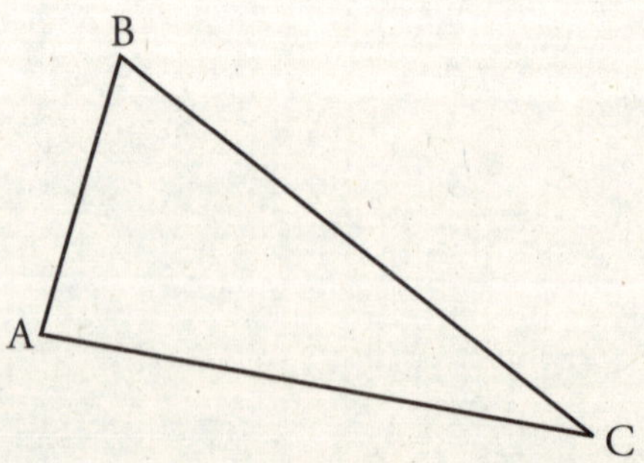

Types of Triangles

If no two sides of a triangle are the same length, then it is a scalene triangle. Naturally, properties of triangles start to get more interesting when some or all of the sides have the same length.

- **Isosceles Triangles:** If two sides of a triangle have the same length, the triangle is isosceles. The relationship between sides and angles still goes; if two sides of a triangle are the same length, then the angles opposite those sides have the same measure.
- **Equilateral Triangles:** Equilateral triangles have three equal sides and three equal angles. Because the sum of the angle measures is 180°, each angle in an equilateral triangle measures $\frac{180}{3}$, or 60°.

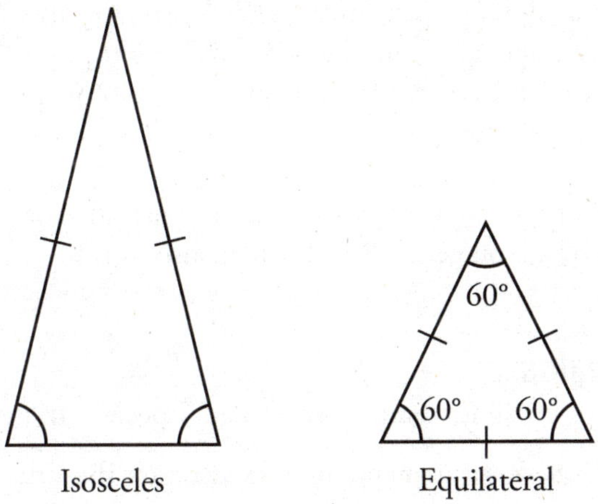

Right Triangles

Right triangles contain exactly one right angle and two acute angles. The perpendicular sides are called legs, and the longest side (which is opposite the right angle) is called the hypotenuse.

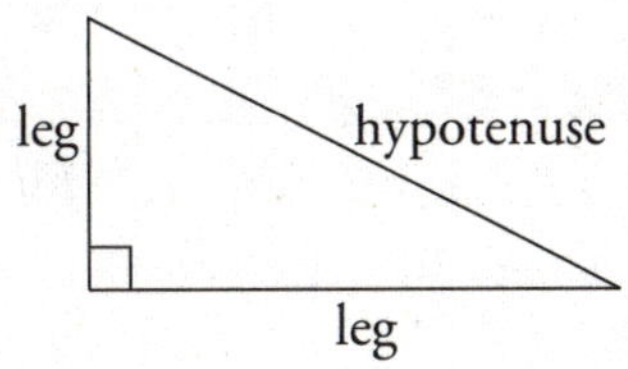

The Pythagorean Theorem

Trigger: Need to know the side of a right triangle.

Response: Write down $a^2 + b^2 = c^2$ and plug in the two sides you know.

Whenever you know the measure of two sides of a right triangle, you can find the length of the third side by using the Pythagorean theorem.

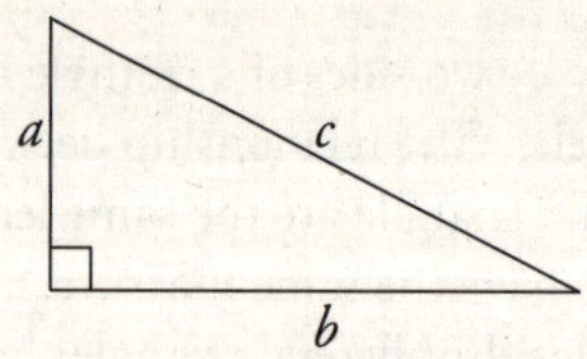

$$a^2 + b^2 = c^2$$

Most of the time, one of the measures of a right triangle is irrational and in the form of a square root. Any set of three integers that works in the Pythagorean theorem is called a "Pythagorean triple," and they're very useful to know for the GRE because they come up often. The most common triple is "3 : 4 : 5" (because $3^2 + 4^2 = 5^2$), but the other three worth memorizing are "5 : 12 : 13," "7 : 24 : 25," and "8 : 15 : 17."

All multiples of Pythagorean triplets also work in the Pythagorean theorem. If you multiply 3 : 4 : 5 by 2, you get 6 : 8 : 10, which also works.

Special Triangles

Two specific types of right triangles are called "special" triangles because their angles and sides have measurements in a fixed ratio. The first, an isosceles right triangle, is also referred to as a 45°–45°–90° triangle because its angle measures are 45°, 45°, and 90°. The ratio of its sides is $1 : 1 : \sqrt{2}$. The second is a 30°–60°–90° triangle, which has sides in a ratio of $1 : \sqrt{3} : 2$.

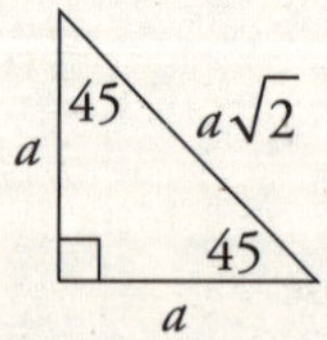

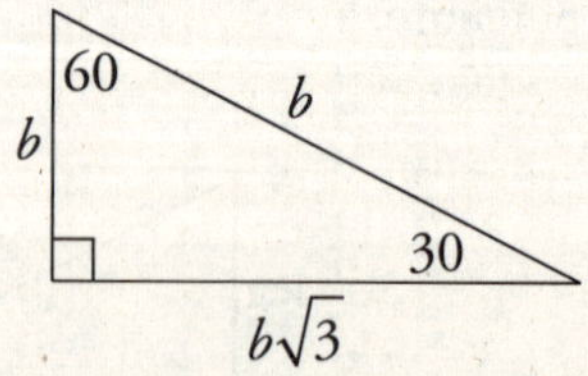

ETS likes to use special triangles because they can confuse test takers into thinking they don't have enough information to answer a question. The fact is, though, if you know the length of one side of a special triangle, you can use the ratios to find the lengths of the other two sides.

Question 18 of 20

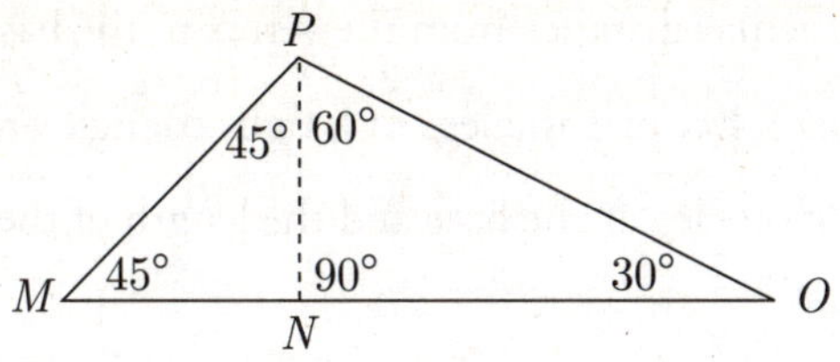

In the diagram above, if the height of the triangle is 4 inches, what is the approximate perimeter, in inches, of ΔMOP?

- ○ $8 + 4\sqrt{2} + 4\sqrt{3}$
- ○ $12 + 8\sqrt{3}$
- ○ $12 + 4\sqrt{2} + 4\sqrt{3}$
- ○ $20 + 8\sqrt{3}$
- ○ $20 + 4\sqrt{2} + 4\sqrt{3}$

Here's How to Crack It

The figure consists of two special triangles; consider the 45°–45°–90° triangle on the left. If $PN = 4$, then $MN = 4$ and $MP = 4\sqrt{2}$. The height PN also helps you find the lengths of the 30°–60°–90° triangle on the right. Because PN is the short side, the hypotenuse PO is twice as long, or 8 inches long. The other side, NO, measures $4\sqrt{3}$. The perimeter of ΔMOP is therefore $4 + 4\sqrt{2} + 8 + 4\sqrt{3}$, which is closest to 24.59. The answer is (C).

Trigger: Triangle question contains the word "area."

Response: Write down $A = \frac{1}{2}bh$ and plug in what you know. The height will always be perpendicular to the base.

Area

The formula for the area of a triangle is $A = \frac{1}{2}bh$, where b is the length of the base and h is the perpendicular distance from the vertex to the base (also known as the height, or the altitude). Because the legs of a right triangle are perpendicular, you can use the length of one leg as the base and the length of the other as the height.

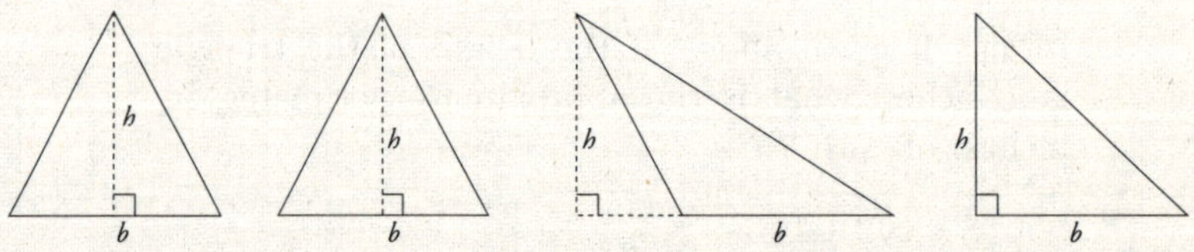

In the diagram above, each of the triangles has the same base and a height of the same measure. Therefore, each triangle has the same area.

Triangle Quick Quiz

Question 1 of 8

If the measures of two angles of ΔABC are 50° and 65°, what is the measure of the third angle?

- ◯ 15°
- ◯ 50°
- ◯ 65°
- ◯ 115°
- ◯ 165°

Question 2 of 8

If isosceles ΔDEF has sides of 11.5 and 13.7, what could the perimeter of the triangle be?

*Indicate **all** possible values.*

- ☐ 2.1
- ☐ 12.0
- ☐ 25.2
- ☐ 36.7
- ☐ 50.4

Question 3 of 8

If the lengths of the legs of a right triangle are 15 and 36, what is the measure of the hypotenuse?

- ○ 21
- ○ 39
- ○ 41
- ○ 48
- ○ 51

Question 4 of 8

The base of a triangle is twice as long as its height, which is 5 cm. What is the area, in square centimeters, of the triangle?

- ○ 6.25
- ○ 10
- ○ 12.5
- ○ 25
- ○ 50

Question 5 of 8

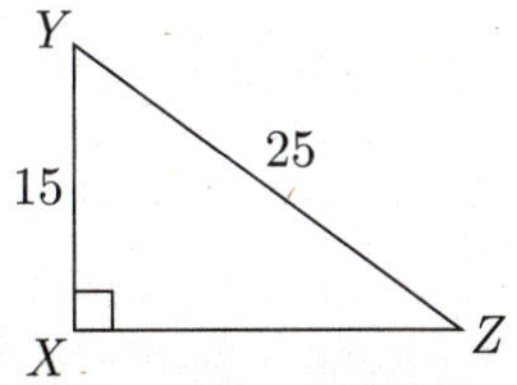

What is the perimeter of triangle XYZ ?

Question 6 of 8

On his way home from shopping, Mike must travel due south for 5 miles, then due east for another 12 miles to reach his house. If Mike could travel in a straight line from the store to his house, how many miles would he save?

- ○ 1
- ○ 4
- ○ 8
- ○ 13
- ○ 17

Question 7 of 8

If the lengths of all 3 sides of triangle *RST* are distinct, single-digit prime numbers, then which of the following could be the perimeter of triangle *RST* ?

- ○ 6
- ○ 10
- ○ 12
- ○ 14
- ○ 15
- ○ 19

Question 8 of 8

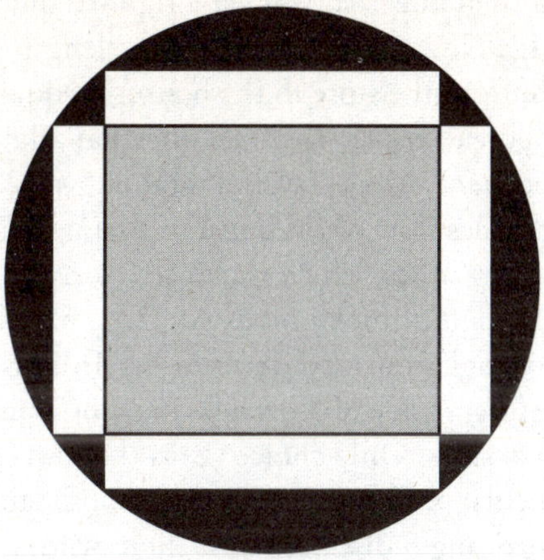

In the above figure, two identical and perpendicular rectangles are inscribed in a circle.

Quantity A	Quantity B
The area of the black region	The area of the gray region

- ○ Quantity A is greater.
- ○ Quantity B is greater.
- ○ The two quantities are equal.
- ○ The relationship cannot be determined from the information given.

Explanations for Triangle Quick Quiz

1. The first two measures are 50° and 65°, so their total measure is 115°. The third angle must measure 180 – 115, or 65°. The answer is (C).
2. The third side of the triangle must be less than the sum of the other two sides or more than the difference. That means that the third side of this triangle, x, is $2.2 < x < 25.2$, which means that the perimeter is $27.4 < P < 50.4$. The answer is (D).
3. You can use the Pythagorean theorem ($225 + 1296 = x^2$), but you might also recognize the Pythagorean triplet 5 : 12 : 13 increased by a multiple of 3. Because $5 \times 3 = 15$ and $12 \times 3 = 36$, then the hypotenuse measures 13×3, or 39. The answer is (B).
4. If the height is 5 cm, then the base is twice as long, or 10 cm. The area is therefore $\frac{1}{2}(5)(10)$, or 25 cm^2. The answer is (D).
5. Using the Pythagorean Theorem, you can find the third side. $25^2 - 15^2 = 400$, and the square root of 400 is 20. The perimeter equals 15 + 20 + 25, or 60. You can also find the third side based on the 3-4-5 Pythagorean triple (just multiply the whole ratio by 5). The answer is 60.

6. Draw out the right triangle described in the question. By traveling due south and then due east, you get a right triangle with legs of length 5 and 12. Now you can solve for the length of the hypotenuse, or the distance from the store to the house. It equals 13 from the 5-12-13 Pythagorean triple. The final question asks for how many miles would be *saved*. Mike travels a total of 5 + 12 = 17 miles as opposed to 13 miles if he could travel in a straight line. 17 – 13 = 4, so he would save 4 miles, which is choice (B).
7. The only single-digit prime numbers are 2, 3, 5, and 7; remember, though, that triangles must conform to the Third Side Rule, which states that the largest side of a triangle must be smaller than the sum of the other two sides. Only choice (E) is the sum of 3 sides that meet both requirements: 3, 5, and 7. Choices (B), (C), and (D) can also be reached by adding 3 distinct single digit primes—2, 3, and 5, 2, 3, and 7, and 2, 5, and 7, respectively—but all violate the Third Side Rule. If you selected choice (A) or choice (F), remember to only use *distinct* values for the sides. The answer is (E).
8. Don't rely on the figure to indicate the dimensions of the rectangles. One extreme would be the rectangles being squares on top of each other, and the other extreme would be the rectangles being so long and skinny that they almost become lines and make the gray area a tiny square. Redrawing and comparing the two columns at those extremes is a good starting point. Draw a diagonal of a rectangle to form two right triangles, the hypotenuse of which is the diameter of the circle. Then use the Pythagorean theorem to relate various lengths within the figure. Starting with the rectangles being squares, the Pythagorean theorem yields: $2 \times s^2 = \text{hypotenuse}^2$. Plug in a number, such as $s = 2$, which makes the area of the gray square 4, and the hypotenuse = $\sqrt{8}$ or $2\sqrt{2}$. The radius of the circle is half the hypotenuse, or $\sqrt{2}$, which makes the area of the circle 2π. Since the two rectangles are, in this extreme case, squares that overlap completely, the area of either square is the gray area, Quantity B, which is 4. The area of the black region is the area of the circle – the area of the square, or $2\pi - 4 = 2.28$. So in that case, Quantity B is greater. In the other extreme, in which the rectangles become very long and skinny, the gray region approaches zero, while the black area gets very large; the black area, Quantity A, is greater. Since you have one situation in which Quantity A is greater and another in which Quantity B is greater, choice (D) is the correct answer.

How Long Is the Third Side?

If you know the lengths of two sides of a triangle, you can use a simple formula to determine how long and how short the third side could possibly be.

> If the lengths of two sides of a triangle are x and y, respectively, the length of the third side must be less than $x + y$ and greater than $|x - y|$.

Question 17 of 20

The townships of Addington and Bordenview are 65 miles apart, and Clearwater is 40 miles from Bordenview. If the three towns are not connected by a straight line, how many miles apart could Addington and Clearwater be?

- ○ 15
- ○ 25
- ○ 35
- ○ 105
- ○ 125

Here's How to Crack It

Here we have a Multiple Choice, Multiple Answer question of the range variety. This is a "third-side" problem disguised as a word problem about three towns. Because the towns do *not* lie along a straight line, they form a triangle; one side is 65 miles long, the other is 40 miles long. Therefore, the third side (the length between towns A and C) must be greater than 65 – 40, or 25, and less than 65 + 40, or 105. (Remember, the distance has to be *greater than* 25, so it can't be *equal* to 25, nor can it be equal to 105.) Therefore, the correct answer is (C).

QUADRILATERALS

A quadrilateral is any figure that has four sides, and the same types of quadrilaterals—parallelograms, rectangles, and squares—show up over and over again on the GRE. Regardless of their shape or size, however, one thing is true of all four-sided figures: They can be divided into two triangles. From this, we can determine a couple of things:

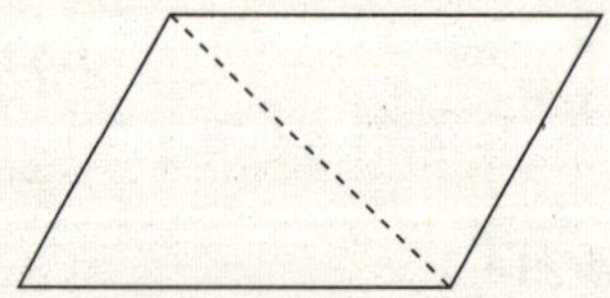

Degrees: Because every quadrilateral can be divided into 2 triangles, all quadrilaterals obey what we call the **Rule of 360:** There are 180 degrees in a triangle, so there are 2 × 180, or 360, degrees in every quadrilateral.

Area: The area of a triangle is $\frac{1}{2}bh$, so the area of a parallelogram is $2 \times \frac{1}{2}bh$, or *bh*.

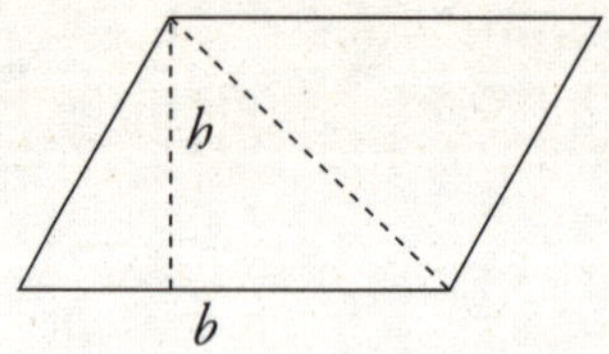

Properties, Area, and Perimeter

Quadrilaterals are often referred to as a "family" because they share lots of characteristics. For example, every rectangle is a parallelogram, so rectangles have every characteristic that a parallelogram has, and they also happen to have four right angles. Ditto for a square, which is just a rectangle that has four equal sides.

Here is a handy chart to help you keep track of all the various properties and the formulas for area and perimeter.

Quadrilateral	Properties	Area	Perimeter
Parallelogram 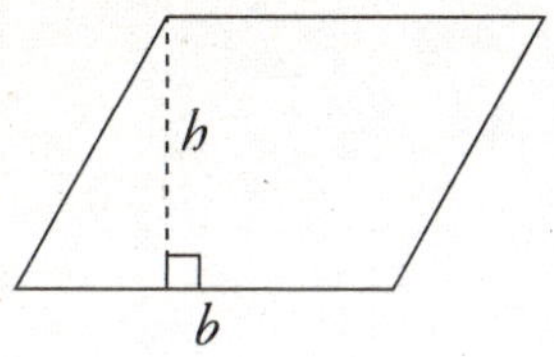	Opposite sides are parallel Opposite sides have the same length Opposite angles have the same measure Diagonals are not the same length (unless it's a rectangle)	$A = b \times h$	Sum of the sides
Rectangle 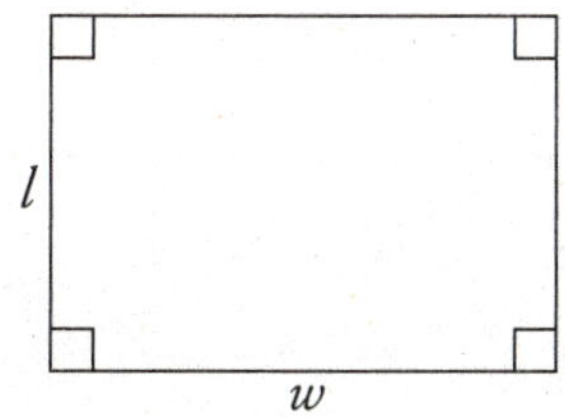	All the properties of parallelograms, plus: Diagonals have the same length All angles have the same measure (90°)	$A = l \times w$	$P = 2l + 2w$
Square 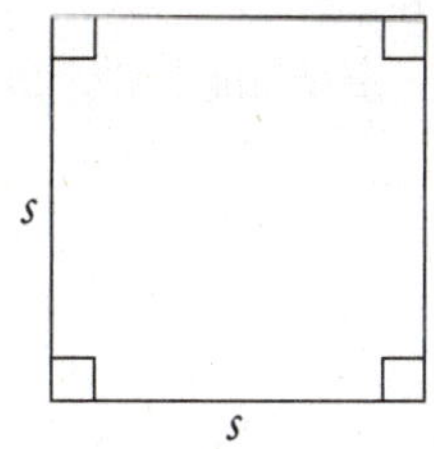	All the properties of rectangles, plus: All sides have the same length	$A = s^2$	$P = 4s$

Trigger: Problem with parallelogram, rectangle, or square contains the word "area."

Response: Write down the area formula and plug in information.

Trigger: Problem mentions "perimeter."

Response: Find the length of each side, and add up all sides.

Quadrilateral Quick Quiz

Question 1 of 5

If the measures of two angles in a quadrilateral are 70° and 130° and the remaining two angles have the same measure, what is the measure of each of them?

- ○ 60°
- ○ 80°
- ○ 100°
- ○ 160°
- ○ 200°

Question 2 of 5

If the length of a rectangular garden is five times its width, and the perimeter of the garden is 36 feet, what is the garden's width?

- ○ 3
- ○ 12
- ○ 15
- ○ 18
- ○ 30

Question 3 of 5

What is the area of the garden in question 2?

[ ]

Click on the answer box, then type in a number. Backspace to erase.

Question 4 of 5

If the base and height of a parallelogram are 10 cm and 15 cm, respectively, what is the area of the parallelogram?

- ○ 25
- ○ 50
- ○ 75
- ○ 150
- ○ It cannot be determined from the information given.

Question 5 of 5

What is the perimeter of the parallelogram in Question 4 ?

- ○ 25
- ○ 50
- ○ 75
- ○ 150
- ○ It cannot be determined from the information given.

Explanations for Quadrilateral Quick Quiz

1. The sum of the given angles is 70° + 130°, or 200°, so the sum of the remaining angles must be 360° – 200°, or 160°. These angles are the same size, so they must each measure $\frac{160}{2}$, or 80°. The answer is (B).
2. If the width of the garden is w, then the length is five times that, or $5w$. Plug these into the formula for perimeter ($2l + 2w$) and solve: $2(5w) + 2w = 36$, so $12w = 36$ and $w = 3$. The garden is 3 feet wide. The answer is (A).
3. The width is three and the length is 5 × 3, or 15. The area is equal to the length times the width, or 15 × 3. The garden's area is 45 square feet.
4. The area of a parallelogram is base × height, so the parallelogram's area is 10 × 15, or 150 cm^2. The answer is (D).
5. You know the base of the parallelogram, but to find the perimeter you need to know the length of the other two sides. There's no way to determine that length from the information given, so the answer is (E).

CIRCLES

A circle represents all the points that are a fixed distance away from a certain point (called the center). The fixed distance from the center to the edge is the radius, and all radii are equal in length. When a radius is rotated 360° around the center, the circumference (the perimeter of the circle) is formed; any segment connecting two points on the circumference is called a chord. The diameter is the longest chord that can be drawn on a circle; it goes through the center and is twice as long as the radius.

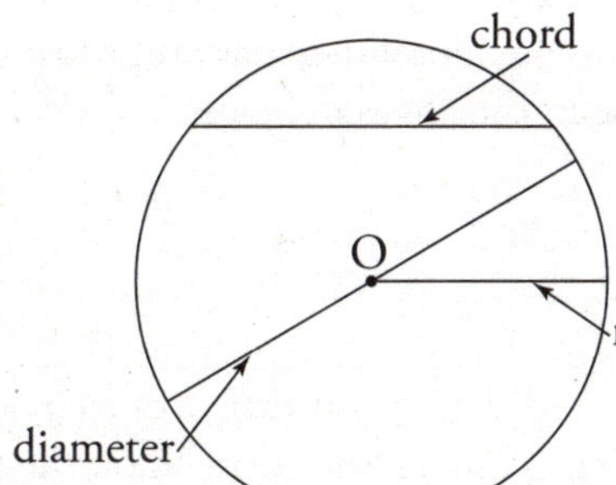

O is the center of the circle.

Area and Circumference

Circles are wondrous things, because they gave us π. One day, a Greek mathematician with a lot of time on his hands began measuring circumferences (C) of circles and dividing those distances by the diameters (d), and he kept getting the same number: 3.141592… He thought this was pretty cool, but also hard to remember, so he renamed it "*p*." He was Greek, though, so he used the Greek letter *p*, which is π.

Trigger: Circle problem contains the word "circumference."

Response: Write $C = 2\pi r$ or $C = \pi d$ on your scratch paper.

Trigger: Circle problem contains the word "area."

Response: Write $A = \pi r^2$ on your scratch paper.

From this discovery we find that $\frac{C}{d} = \pi$, and this can be rewritten as $C = \pi d$. This is the most common formula for finding the circumference of a circle. A diameter is twice as long as a radius ($d = 2r$), so you can also write the formula as $C = 2\pi r$. The formula for the area of a circle is $A = \pi r^2$.

- Area of a circle = πr^2
- Circumference of a circle = $2\pi r$

Notice that the radius, r, is in both of those formulas? The radius is the most important part of a circle to know. Once you know the radius, you can easily find the diameter, circumference, or the area. So if you're ever stuck on a circle question, find the radius.

One quick note about π. Although it is true that $\pi \approx 3.1415$ (and so on and so on), you won't have to use that too often on the GRE. Don't worry about memorizing π beyond the tenths digit: It's 3.1. Even that is more precise than the GRE typically requires. Most answers are going to be in terms of π, which means that the GRE is much more likely to have 5π as an answer choice than it is to have 15.707. So don't multiply out π unless you absolutely have to. Most of the time, each little π will either cancel out or be in the answer choices.

Sectors and Arcs

An arc is a measurement around the circumference of a circle, and a sector is a partial measurement of the area of a circle. Both depend on the measure of the central angle, which has its vertex on the center of the circle.

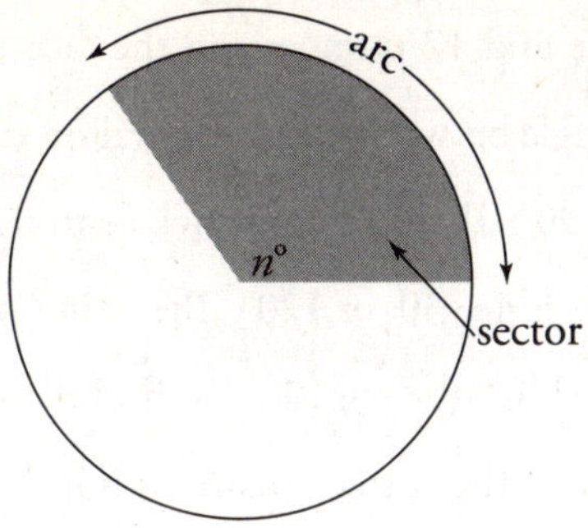

We will figure out the length of an arc or the area of a sector by comparing it to the entire circle. If the central angle is 180°, then the arc must be half of the circumference, because 180° is half of the total 360° in the central angle of a circle. A sector made up of a 90° angle must be $\frac{1}{4}$ of the total area, because 90° is one fourth of 360°.

$$\frac{\textit{angle}}{360°} = \frac{\textit{arc}}{\textit{circumference}} = \frac{\textit{sector}}{\textit{area}}$$

Question 14 of 20

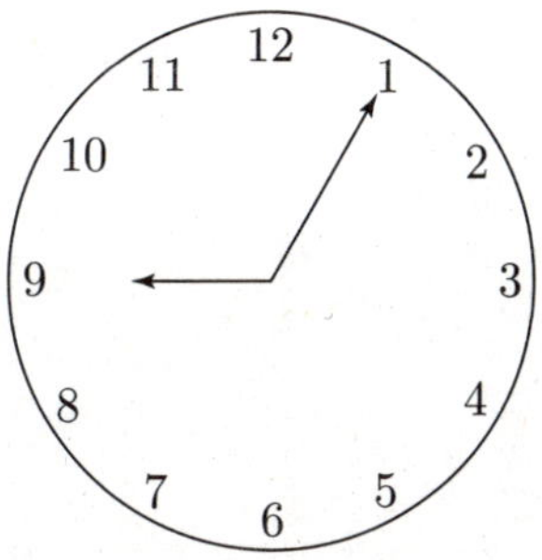

In the diagram above, the clock shows that it is five minutes after 9 o'clock. If the radius of the clock is 9 inches long, what is the area of the sector intercepted by the hands of the clock?

- ○ 6π
- ○ 13.5π
- ○ 27π
- ○ 81π
- ○ 54π

Here's How to Crack It

There are 360° in a circle and 12 numbers on the face of a clock. Therefore, the measure of the central angle between each numeral on the clock (say, between the 12 and the 1) is $\frac{360}{12}$, or 30°. There are four such central angles between the 9 and the 1, so the central angle is 4×30, or 120°. The radius of the circle is 9 inches, so the area of the whole clock is $\pi(9)^2$, or 81π. To find the area of the sector, use the formula: $81\pi \times \frac{120}{360} = 27\pi$. The correct answer is (C).

Circle Quick Quiz

Question 1 of 4

What is the diameter of a circle that has a circumference of 6π inches?

- ○ 2
- ○ 3
- ○ 4
- ○ 6
- ○ 12

Question 2 of 4

What is the area of the circle in question 1 ?

- ○ 3π
- ○ 6π
- ○ 9π
- ○ 12π
- ○ 36π

Question 3 of 4

A cherry pie with a radius of 8 inches is cut into six equal pieces. What is the area, in square inches, of each slice?

- ○ $\frac{8\pi}{3}$
- ○ $\frac{16\pi}{3}$
- ○ $\frac{32\pi}{3}$
- ○ $\frac{64\pi}{3}$
- ○ $\frac{128\pi}{3}$

Question 4 of 4

A cherry pie with a radius of 8 inches is cut into six equal pieces. What is the measure of the central angle of each piece of pie?

- ○ 15°
- ○ 30°
- ○ 60°
- ○ 90°
- ○ 120°

Explanations for Circle Quick Quiz

1. Because $C = \pi d$, the diameter of a circle with a circumference of 6π inches is 6 inches. The answer is (D).

2. The diameter of the circle is 6, so the radius is half that, or 3. The area of the circle is $\pi(3)^2$, or 9π square inches. The answer is (C).

3. The area of the pie is πr^2, or 64π, so each slice has an area of $\frac{64\pi}{6}$ square inches. This converts to $\frac{32\pi}{3}$. The answer is (C).

4. There are 360° in a circle, so the measure of each central angles is $\frac{360}{6}$, or 60°. The answer is (C).

THE COORDINATE PLANE

You might find a smattering of questions about the *x-y* plane, otherwise known as the Cartesian or coordinate plane on the GRE. The primary skill you'll need to possess in order to answer these questions is being able to plot points by using the calibrations on the *x*- and *y*-axes. These two lines divide the space into four quadrants.

When plotting the point (x, y) on the coordinate plane

- start at the point (0, 0), which is also known as "the origin"
- move x units to the right (if x is positive) or left (if x is negative)
- move y units up (if y is positive) of down (if y is negative)

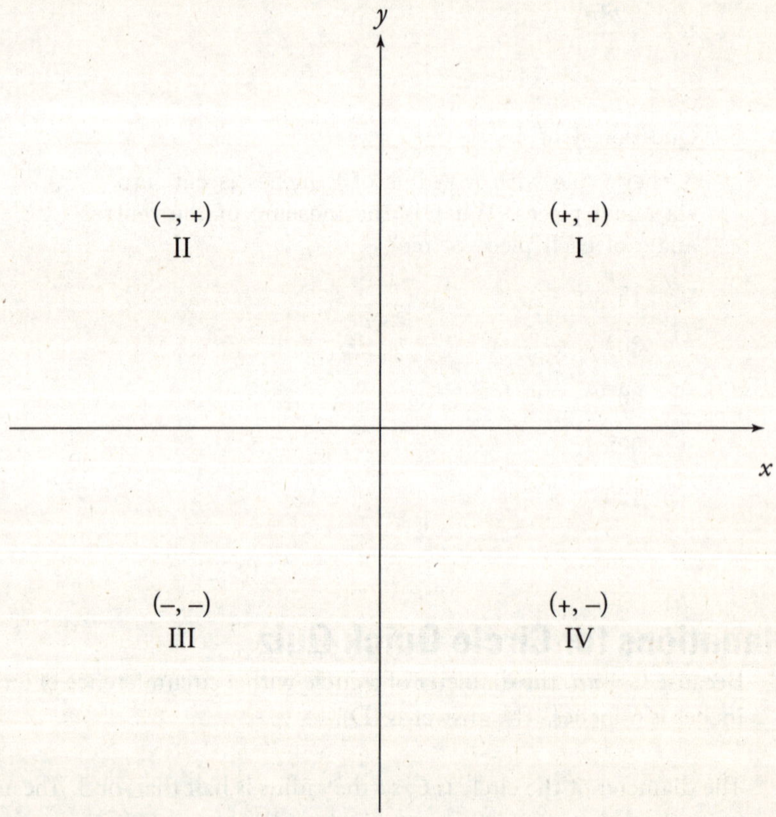

When you plot points on the Cartesian plane, you'll most likely be asked to (1) find the distance between two of them, (2) find the slope of the line that connects the two points, or (3) find the equation of the line they define.

Distance in the Coordinate Plane

If you need to find the distance between two points in the coordinate plane, draw a right triangle. The hypotenuse of the triangle is the distance between the two points, and the legs are the differences in the *x*- and *y*-coordinates of the two points.

Question 11 of 20

If point A is at (2, 3) in the rectangular coordinate plane, and point B is at (–3, 15), what is the length of line segment $\overline{AB}$?

Here's How to Crack It

Start by drawing a simple coordinate plane on your scratch paper. You don't need to mark out each and every tick mark; this is just to get a rough idea of where our points are. Your drawing will probably look something like this:

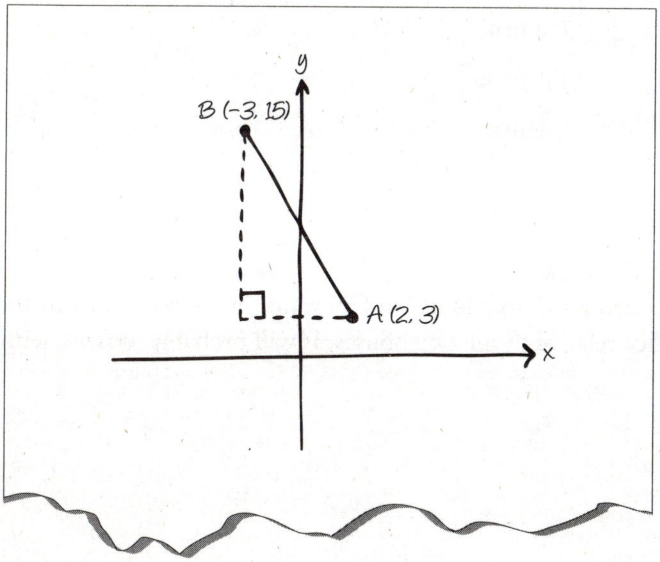

Now let's find the length of each leg. The bottom leg is the distance between our two x-coordinates. From –3 to 2 is a total of 5 units (or, $|-3-2|=5$). Our height is the distance between our two y-coordinates. From 3 to 15 is 12 units (or, $|3-15|=12$). So we have a triangle with sides of length 5 and 12. We can either use Pythagorean theorem to find the hypotenuse, or use the fact that we have a 5 : 12 : 13 triangle, which means that the line is 13 units long.

Many two-dimensional distance problems are really Pythagorean theorem problems in disguise. The GRE will often hide this using questions in which people travel north/south and east/west.

Question 12 of 20

Mark and Kim are at the post office. To get home, Kim walks 3 miles north and 2 miles west. Mark walks 8 miles south and 2 miles east to get home. What is the approximate straight-line distance between Kim's house and Mark's house?

- ○ 10.1 miles
- ○ 11.7 miles
- ○ 12 miles
- ○ 13.2 miles
- ○ 15 miles

Here's How to Crack It

Let's draw a diagram, first. Nothing fancy, just enough to show us the location of the post office relative to the two houses. It will probably look something like this:

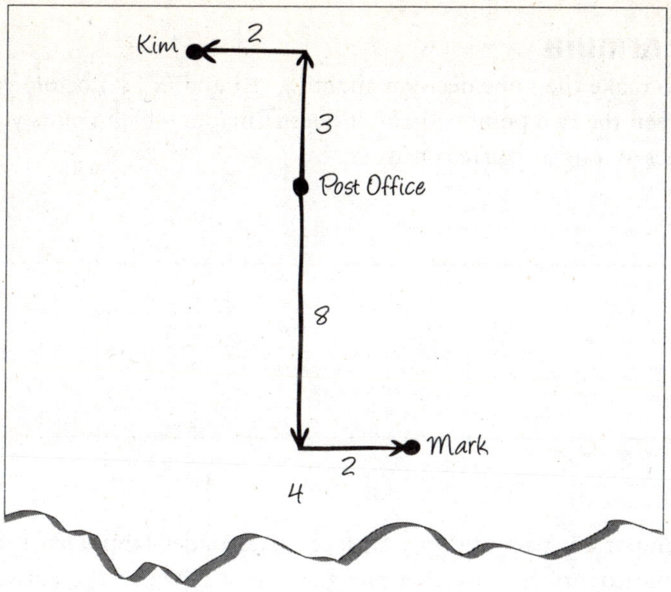

Now that we've got that drawn, let's make our triangle. The trick here is that we can now completely ignore the post office. We just want to know the distance from Kim's house to Mark's house, so those will be the two points of our triangle, like so:

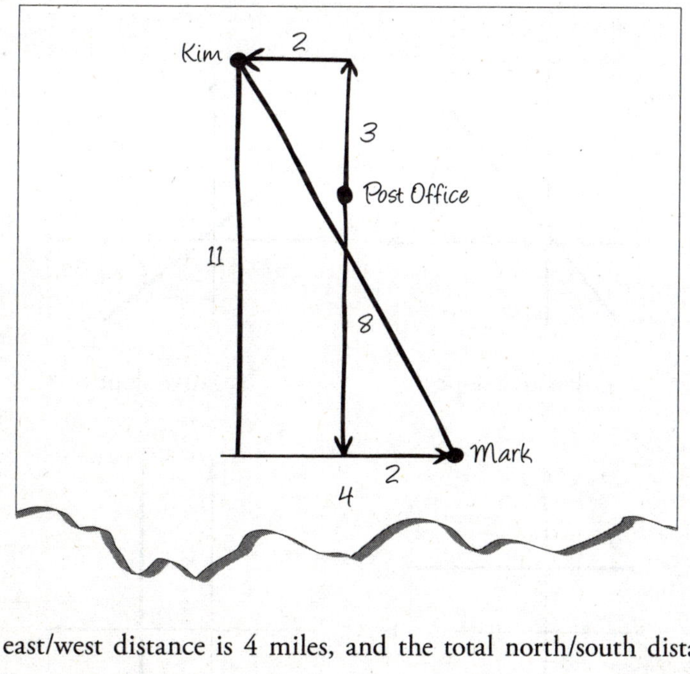

The total east/west distance is 4 miles, and the total north/south distance is 11 miles. Those are the legs of our triangle. We can plug those into the Pythagorean theorem, and we'll get $4^2 + 11^2 = c^2$; $16 + 121 = c^2$; $137 = c^2$; $c = \sqrt{137} \approx 11.7$, answer (B).

Slope Formula

You have to make the same decision about (x_1, y_1) and (x_2, y_2) before you find the slope between the two points. Again, it doesn't matter which points you assign to which values as long as you're consistent.

$$\text{Slope of a line} = \frac{rise}{run} = \frac{y_2 - y_1}{x_2 - x_1}$$

The most important part of slope, however, is to understand what it means. The slope is a measure of how much a line goes up or down in the y (rise) as it goes over in the x (run). In simpler language, the slope measures how slanted the line is. A positive slope means that the line rises up from left to right. A negative slope means that the line goes down from left to right. Notice that we always read the line from left to right, as when reading a sentence. A slope of zero means that as we go over, the line never rises: It just stays a level, flat line. An undefined slope means that the line never runs over; it just goes up and up and up. (The slope is undefined because the difference in x-coordinates for any two points is 0, and we can't divide by 0.)

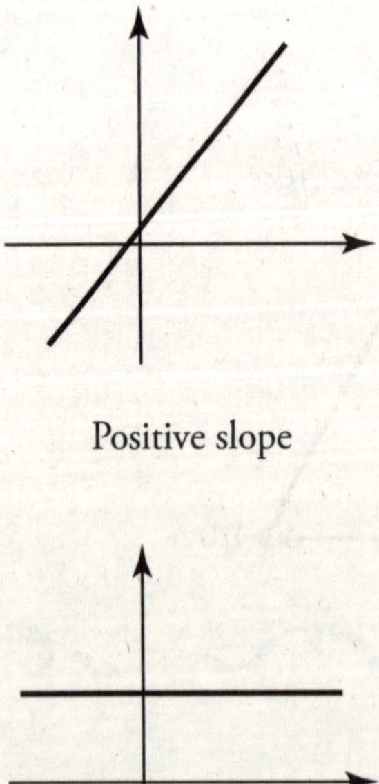

Positive slope

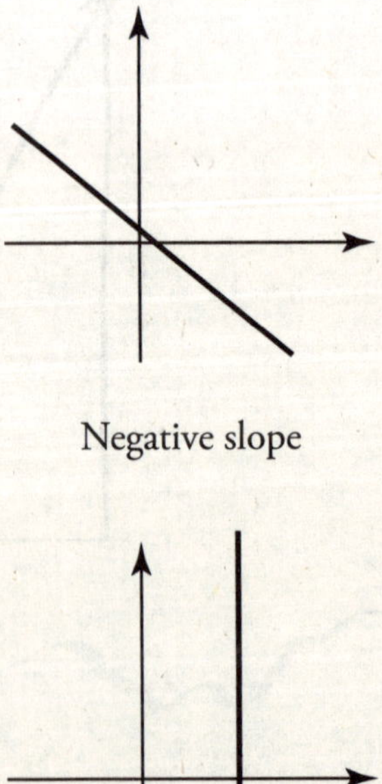

Negative slope

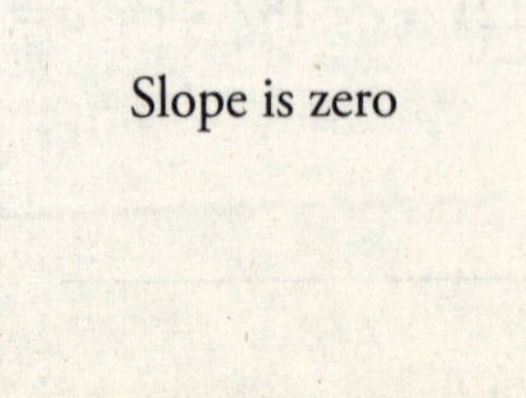

Slope is zero

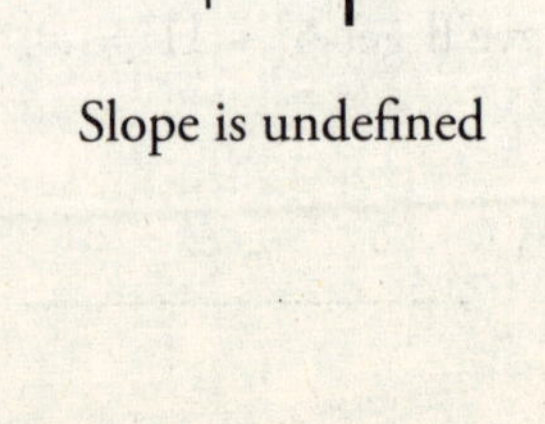

Slope is undefined

Equation of a Line

Any line on the coordinate plane can be represented in the form $y = mx + b$, in which m is the slope of the line and b is the y-intercept. For example, the line $y = 3x - 6$ has a slope of 3 and hits the y-axis at the point (0, –6).

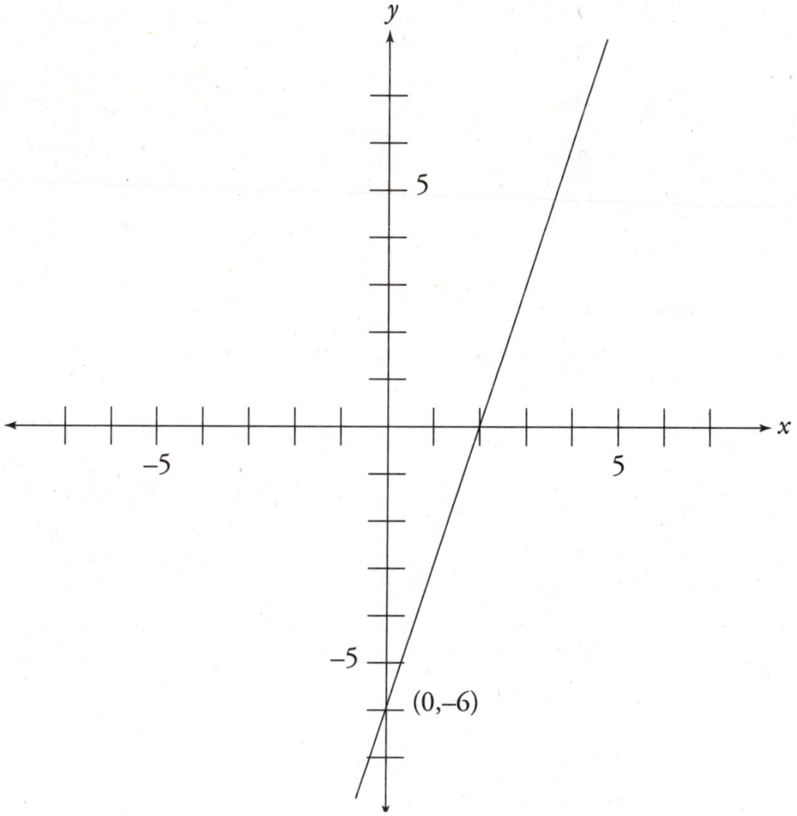

Point-Slope Formula

If you know the slope m of a line and the coordinates of one of the points on that line (x_0, y_0), you can use the point-slope formula to determine the equation of the line: $y - y_0 = m\,(x - x_0)$.

Coordinate Plane Quick Quiz

Question 1 of 5

If a line has a slope of –3 and passes through the point (2, 1), what is the equation of the line?

- ○ $y = -3x + 1$
- ○ $y = -3x + 7$
- ○ $y = -3x + 11$
- ○ $y = 3x + 1$
- ○ $y = 3x + 5$

Question 2 of 5

What is the slope of the line between the points (–4, 3) and (6, –2)?

- ○ -2
- ○ $-\frac{1}{2}$
- ○ $\frac{1}{2}$
- ○ 2
- ○ 4

Question 3 of 5

What is the distance between the points in question 2 ?

- ○ $5\sqrt{2}$
- ○ $5\sqrt{5}$
- ○ $6\sqrt{3}$
- ○ $6\sqrt{5}$
- ○ $7\sqrt{2}$

Question 4 of 5

What is the equation of the line formed by the points in question 2 ?

○ $y = -\frac{1}{2}x + 1$

○ $y = -\frac{1}{2}x + 2$

○ $y = \frac{1}{2}x + 5$

○ $y = \frac{1}{2}x + 7$

○ $y = \frac{1}{2}x + 8$

Question 5 of 5

At what point(s) does the line $y = 2x + 6$ intersect the x- and y-axes?

○ (0, –3)

○ (0, 6)

○ (–6, 0)

○ (–3, 0)

○ (3, 0)

Explanations for Coordinate Plane Quick Quiz

1. You have a point and the slope, so use the point-slope formula: $(y - 1) = -3(x - 2)$. Simplified, this becomes $y - 1 = -3x + 6$. Add 1 to both sides, and your final answer is $y = -3x + 7$. The answer is (B).

2. Use the slope formula: $\frac{-2-3}{6-(-4)} = \frac{-5}{10} = -\frac{1}{2}$. The answer is (B).

3. Draw a triangle on your scratch paper. The bottom of the triangle goes from –4 to 6, so the base is 10 units long. The height of the triangle goes from –2 to 3, so it is 5 units long. Plugging these lengths into the Pythagorean theorem, $a^2 + b^2 = c^2$, gives us $5^2 + 10^2 = c^2$; $25 + 100 = c^2$; $125 = c^2$; $c = \sqrt{125} = \sqrt{25} \times \sqrt{5} = 5\sqrt{5}$, answer (B).

4. You know the slope, and you can choose either point you were given. Use the point-slope formula: $y - (-2) = -\frac{1}{2}(x - 6)$. This simplifies to $y = -\frac{1}{2}x + 1$. The answer is (A).

5. You are given the formula for the line, $y = 2x + 6$. When the line crosses the x-axis, the value of y, at that point, will be 0. Eliminate answer choices (A) and (B). Now plug 0 in for y in the equation for the line and solve for x. $x = -3$, so the point is (–3,0) and the answer is (D).

VOLUME FORMULAS

Volume problems on the GRE are rare and will involve only a select few geometric solids. Formulas for those solids are below, and it's best just to memorize them, just in case a volume problem comes up.

Solid	Volume Formula
Rectangular prism	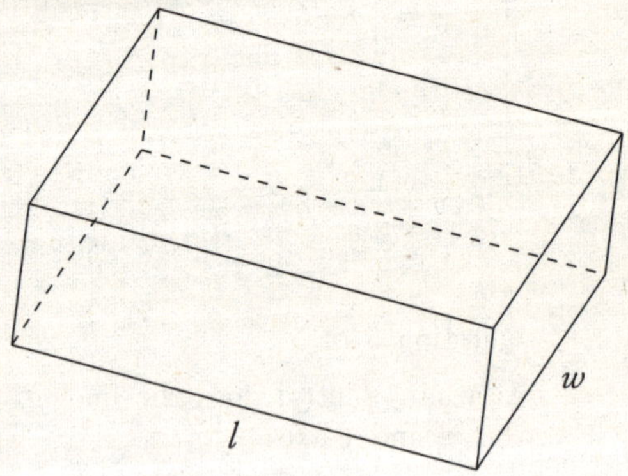 $V = l \times w \times h$
Right circular cylinder	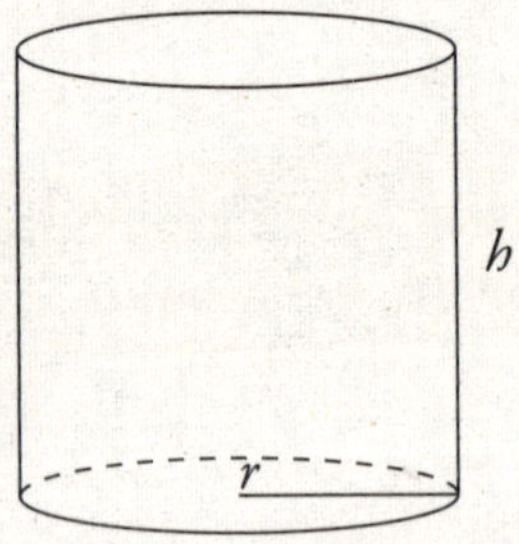 $V = \pi r^2 h$

Space Diagonal

If you know the dimensions of a rectangular prism, you can determine the length of the greatest distance between any two points in that prism. This distance is called the space diagonal, which can be found using what looks like an extension of the Pythagorean theorem.

> If the dimensions of a rectangular prism are a, b, and c, then the space diagonal d can be found using the formula $d^2 = a^2 + b^2 + c^2$.

SURFACE AREA

Predictably, the surface area of a three-dimensional figure is the sum of all of its surfaces. Almost all surface area questions on the GRE will deal with rectangular prisms. Here is the formula for the surface area of a rectangular prism.

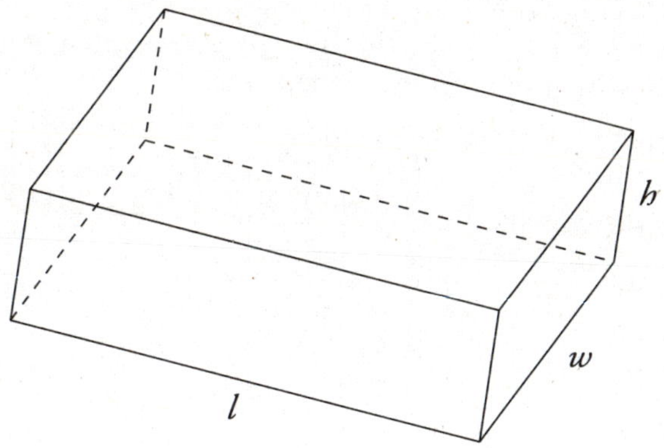

$$SA = 2lw + 2wh + 2lh$$

Geometry Drill

Question 1 of 15

Lines X, Y, and Z intersect to form a triangle. If line X is perpendicular to line Y and line X forms a 30-degree angle with line Z, which of the following is the measure of an angle formed by the intersection of lines Y and Z ?

○ 20°

○ 30°

○ 45°

○ 50°

○ 60°

Question 2 of 15

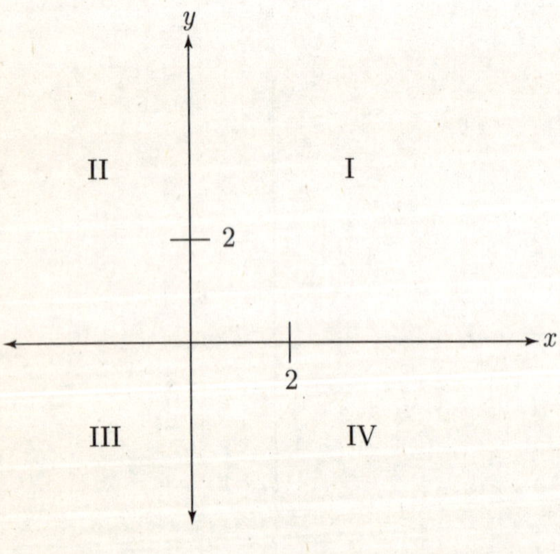

Points (x, 5) and (–6, y), not shown in the figure above, are in Quadrants I and III, respectively. If $xy \neq 0$, in which quadrant is point (x, y)?

○ IV

○ III

○ II

○ I

○ It cannot be determined from the information given.

Question 3 of 15

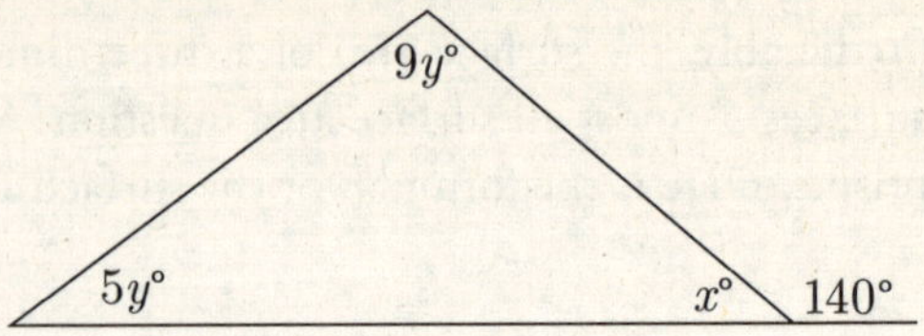

In the triangle above, what is the degree measure of the smallest angle?

○ 10°

○ 40°

○ 45°

○ 50°

○ 60°

Question 4 of 15

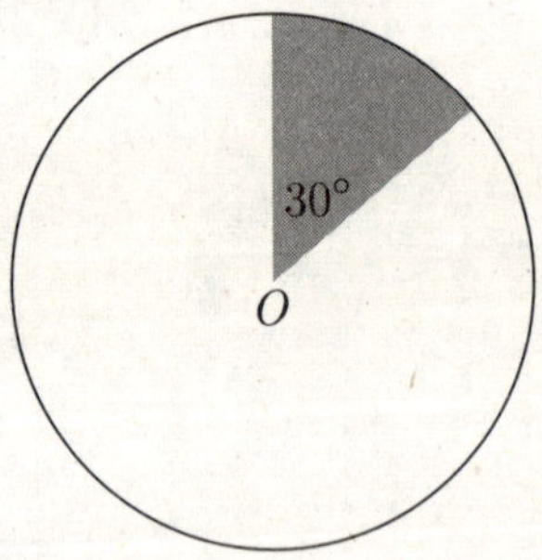

The area of the shaded region is 3π. What is the radius of circle with center O ?

○ 3

○ 6

○ 9

○ 12

○ 36

Question 5 of 15

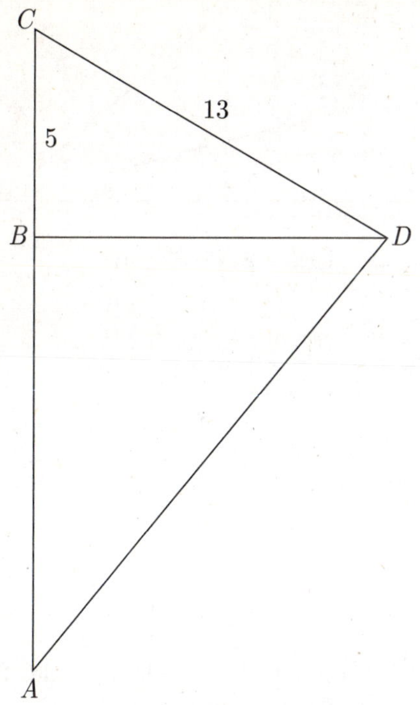

In the figure above, if BD is perpendicular to AC and $AC = 21$, then $AD =$

- ○ 12
- ○ 13
- ○ 20
- ○ 21
- ○ 25

Question 6 of 15

Which of the following expresses the area of a square region in terms of its perimeter p ?

- ○ $\frac{p}{4}$
- ○ $\frac{p^2}{4}$
- ○ $\frac{p}{16}$
- ○ $\frac{p^2}{16}$
- ○ $\left(\frac{4p}{p}\right)^2$

Question 7 of 15

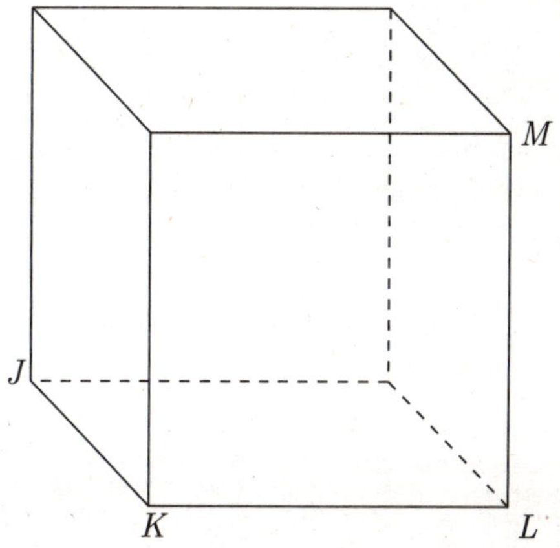

The figure above is a rectangular solid in which $JK = 2$, $KL = 9$, and $LM = 9$. What is the total surface area of the figure?

- ○ 234
- ○ 162
- ○ 134
- ○ 117
- ○ 20

Question 8 of 15

David plans to purchase a triangular fish tank to go in a corner of his rectangular living room. If the clearance on each wall from the corner of the room is 18 inches, and all sides of the tank can touch the walls, then what is the maximum length, in inches, of the front side of the fish tank?

- ○ $16\sqrt{2}$
- ○ 18
- ○ $18\sqrt{2}$
- ○ 36
- ○ $24\sqrt{3}$

Question 9 of 15

The four corners of the face of a cube have coordinates (a, b), (a, d), (c, b) and (c, d). If $a = 2$ and c is an even number between 6 and 11, which of the following could be the surface area of the cube?

- ○ 96
- ○ 150
- ○ 294
- ○ 384
- ○ 600

Click on your choice(s).

Question 10 of 15

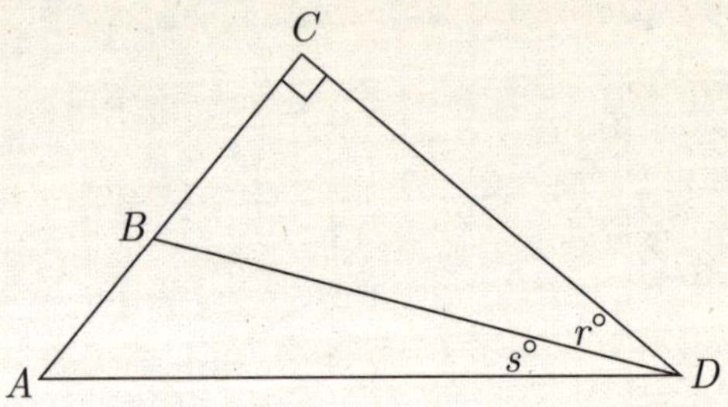

In the drawing above, if $AC = CD$, then $r =$

- ○ $45 - s$
- ○ $90 - s$
- ○ s
- ○ $45 + s$
- ○ $60 + s$

Question 11 of 15

Triangle ABC is an isosceles right triangle. If $AB = AC$, then the area of a square with a side length equal to twice the length of BC is how many times the area of triangle ABC ?

[ ]

Question 12 of 15

Which of the following has the greatest value?:

- ○ The area of a rectangle with length 11 and height 6
- ○ The area of a right triangle with base length 10 and height 10
- ○ The area of a square with diagonal $8\sqrt{2}$
- ○ The area of a circle with a radius of 5
- ○ The area of an equilateral triangle with a side of 12

Question 13 of 15

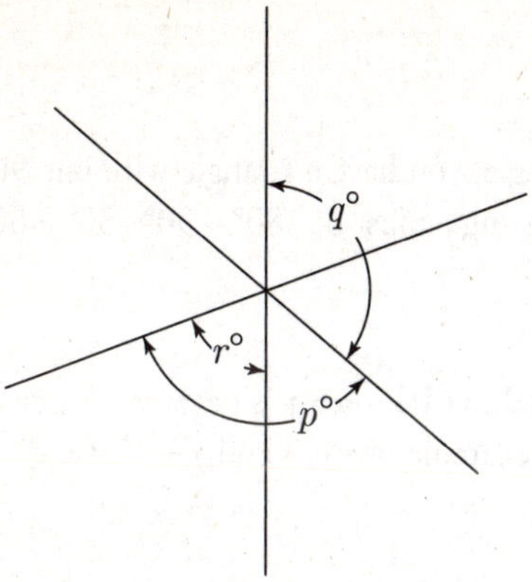

In the figure above, if $q = 130$ degrees and $p = 120$ degrees, then $r =$

- ○ 20°
- ○ 60°
- ○ 70°
- ○ 80°
- ○ 90°

Question 14 of 15

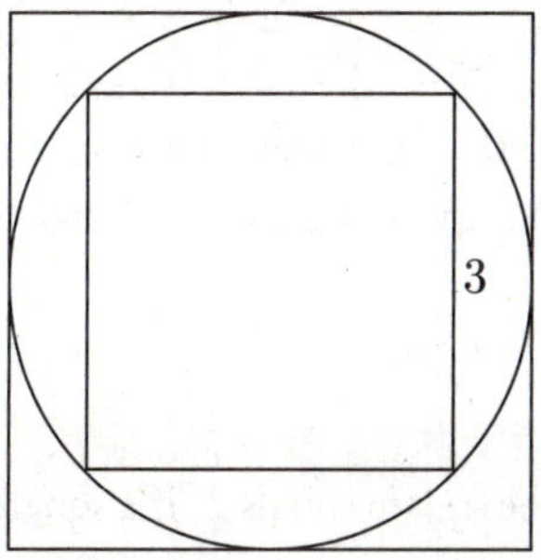

In the figure above, a square with side of length 3 is inscribed in a circle that is inscribed in a square. What is the area of the larger square?

[ ]

Click on the answer box, then type in a number. Backspace to erase.

Question 15 of 15

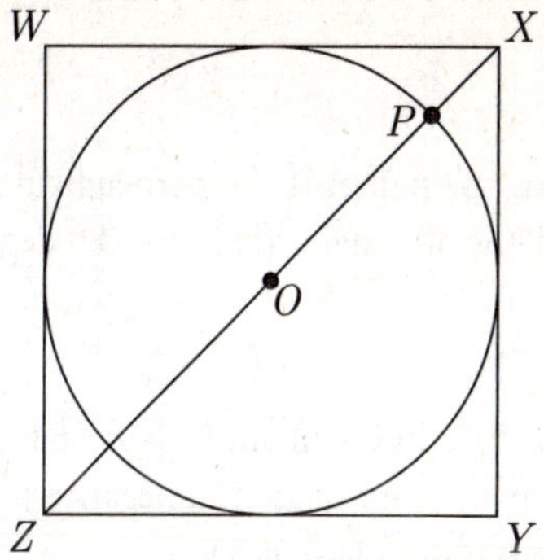

In the figure above, a circle with center O is inscribed in square $WXYZ$. If the circle has radius 3, then $PZ =$

- ○ 6
- ○ $3\sqrt{2}$
- ○ $6 + \sqrt{2}$
- ○ $3 + \sqrt{3}$
- ○ $3 + 3\sqrt{2}$

EXPLANATIONS FOR GEOMETRY DRILL

1. E

Draw the figure. If X is perpendicular to Y, and X and Z form a 30° angle, you have a triangle with one 90° angle and one 30° angle. There are 180 degrees in a triangle, so the remaining angle must be 180° – 90° –30° = 60°.

2. A

If (x, 5) is in Quadrant I, then x has a positive value. If (–6, y) is in Quadrant III, then y is negative. A point with a positive x-coordinate and negative y-coordinate is in Quadrant IV. For example, if $x = 3$ and $y = -5$, $(x, y) = (3, -5)$, which is in Quadrant IV.

3. B

Use the Rule of 180 for a line to find + that the value of x is 180° – 140° = 40°. Next, apply the rule of 180 for a triangle to determine that $9y + 5y = 40°$. Thus, $y = 10$, the angle represented by $5y$ is 50°, and the angle represented by $9y$ is 90°. That angle represented by x, which is 40°, is the smallest angle, making choice (B) the correct answer.

4. B

Because $\frac{30}{360} = \frac{1}{12}$, the shaded region is $\frac{1}{12}$ of the circle. Multiply 3π by 12 to find the area of the entire circle, 36π. Now put the values you know into the formula for the area of the circle: $36\pi = \pi r^2$. Solve for r to find the radius is 6. The answer is (B).

5. C

Since BD is perpendicular to AC, it cuts the large triangle into two right triangles. The small triangle on top, BCD, is the familiar 5-12-13 triangle, so $BD = 12$. Since $AC = 21$ and $CB = 5$, $AB = 16$. Now you have the two short sides of ABD, 12 and 16; use the Pythagorean Theorem or recognize the multiple of the 3-4-5 triangle to find that $AD = 20$.

6. D

Plug in, and let $p = 8$. If the perimeter of the square is 8, then each side of the square equals 2. If a square has side 2, then its area is 4. 4 is the target. Put $p = 8$ into the answer choices. Answer choice (D) is the only one that matches the target of 4: $\frac{8^2}{16} = \frac{64}{16} = 4$.

7. A

To find surface area, add the areas of each face. The rectangle facing outward has sides of $KL = 9$ and $LM = 9$, so it has an area of $9 \times 9 = 81$. The face opposite this front face is identical, and the area of those two faces totals 162. Now look at the face on the left of the solid: It has sides of 2 and 9. The area of this face is $2 \times 9 = 18$, and again, the opposite face is the same, for a total area of 36. Finally, the bottom and top faces are each 2×9, for a total area of 36. Add up the areas of all faces to find the total surface area: 162 + 36 + 36 = 234, choice (A).

8. C

First, draw a rectangle, and then add a line to make one of the corners a triangle. The clearance on each wall is the same, so you have a 45-45-90 right triangle in which the hypotenuse is the front of the tank. The ratio of the sides is $s : s : s\sqrt{2}$; since the length of a short side is 18 inches, the front of the tank must be $18\sqrt{2}$.

9 D

First, draw the face of the cube in the coordinate plane.

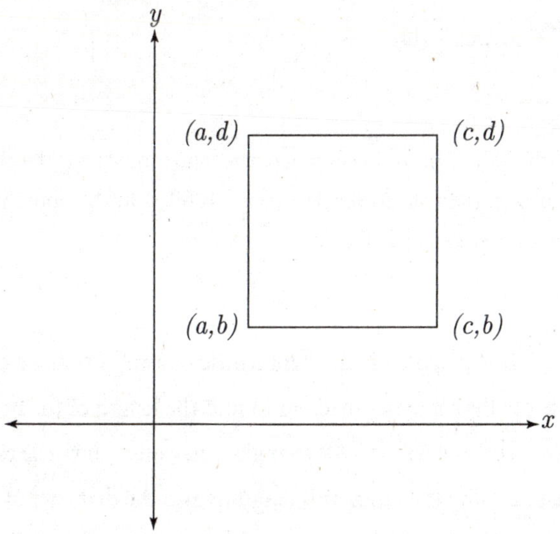

Because $a = 2$ and c could be 8 or 10, one side of the square face of the cube could measure 6 or 8. Now take the possible areas for a single face, found by squaring the side, and multiply by 6 to account for the 6 equal faces of the cube: $6^2 \times 6 = 36 \times 6 = 216$ and $8^2 \times 6 = 64 \times 6 = 384$. Only 384 is an option, so the answer is (D).

10. A

$\Delta\, BCD + \Delta\, ABD$ make the big $\Delta\, ACD$. The problem tells you that $AC = CD$. These are the legs of the big $\Delta\, ACD$. Because $\angle\, ACD$ is a right angle, and we know the two smaller angles of $\Delta\, ACD$ are equal (because the sides are equal), then each of the angles must be 45 degrees. There are variables in the answer choices, so plug in for s and r. If $s = 20$, then $r = 25$. If you plug in 20 in for s, then only answer choice (A) works.

11. 16

First, draw triangle ABC, and plug in sides; *isosceles right triangle* is another way of saying 45-45-90 triangle, so make the short sides 10 and the long side, or BC, $10\sqrt{2}$. Now the area of triangle ABC is 50. Since $BC = 10\sqrt{2}$, the square has a side of $20\sqrt{2}$, and an area of 800. To find how many times the area of the triangle this represents, calculate $800 \div 50 = 16$ times the area of triangle ABC.

12. D

Unfortunately you will have to calculate them all. (A) is length by width by height, which equals 66. (B) is $\frac{1}{2}bh$, which is 50. Eliminate (B). The square in (C) has a side length of 8 because the diagonal of a square is the hypotenuse of a right isosceles triangle with a ratio of sides of $x : x : x\sqrt{2}$, so the area of the square is 64^2. Eliminate (C). (D) has an area of πr^2 which in this case is 25π, which will be something slightly north of 75. Eliminate (A). For (E) the height of an equilateral triangle is the middle side of a 30 : 60 : 90 triangle which has a ratio of sides of $x : x\sqrt{3} : 2x$. In this case the area of the triangle, $\frac{1}{2}bh$, is $\frac{1}{2}(12)(6\sqrt{3})$. $\sqrt{3}$ is less than two. 36 × 2 is still less than answer choice D, so you can eliminate (E). The answer is (D).

13. C

If $p = 120°$, then by the rule of 180 for lines, the unnamed angle in the lower half of the figure must be $180° - 120° = 60°$. Because this unnamed angle is also the lower part of $\angle q$, then the lower part of $\angle q$ is 60° and the upper part is 70° $q = 130°$. Because $\angle r$ is opposite the upper portion of $\angle q$, r is also 70°.

14. E

Start by drawing the diagonal of the smaller square. This diagonal is the hypotenuse of a right triangle created by the diagonal and two sides of the small square. Use the Pythagorean theorem to find the length of the hypotenuse: $c = \sqrt{3^2 + 3^2} = 3\sqrt{2}$. Alternatively, if you recognized this as a 45°-45°-90° triangle, you would know the ratio of the sides is $a : a : a\sqrt{2}$, and you could find the hypotenuse using the ratio. This length is also the diameter of the circle, and the diameter of the circle is equal to the length of the sides of the big square. Use the formula for the area of a square, $A = s^2$, to find the area of the larger square. $A = \left(3\sqrt{2}\right)^2 = 18$. The answer is (E).

15. E

Draw a diameter across the circle. The diameter is the same length as a side of the square. The radius of the circle is 3, the diameter is 6, and the side of the square is also 6. XZ forms a diagonal of the square, and it divides the square into two 45°-45°-90° right triangles. Because you know the side of the square is 6, you can use the 45°-45°-90° ratio ($a : a : a\sqrt{2}$) to find that the length of the diagonal is $6\sqrt{2}$. Finally, PZ can be divided into two pieces, one from O to P and one from O to Z. Line segment OP is equal to the radius of the circle, 3. Line segment OZ is equal to half the diagonal of the square, or half of $6\sqrt{2}$. The length of PZ must be the two added together, or $3 + 3\sqrt{2}$, choice (E).

Chapter 8
The Rest of the Story

TYING UP LOOSE ENDS

So here it is, the last chapter about quantitative topics on the GRE. Most of the material we review in this chapter will probably appear a little less frequently on the GRE, and most of it deals with data analysis. There are a lot of formulas in this chapter that you'd do well to memorize, because it will save you a considerable amount of time when you sit down to take the test. Let's begin.

FUNCTIONS

Sometimes the GRE will make up a mathematical operator. You will see some weird shape, which you've never seen in a math problem, and be asked to solve a problem using this weird shape. This is simply the GRE trying to frighten you.

These questions are asking about functions. A function is simply a set of directions. For instance, think of the word "chop" in a recipe. The word chop is actually telling you to do many things: take out a cutting board, rinse the vegetable or fruit to be chopped, take out a knife, place the vegetable on the cutting board, et cetera et cetera. If a recipe says "Chop 3 carrots," then you must do all those things that chopping entails, but using carrots. If the recipe says "Chop 2 stalks of celery," then you must do all of those chopping things, but with celery.

A function is the same way. It is a set of rules, and the GRE will ask you to perform those rules on a certain number. If the GRE invents some function, for instance that $\blacklozenge x = 4x + 2$, and then asks what $\blacklozenge 5$ is, then figure out what you need to replace. Our original function was $\blacklozenge x$, but notice how with $\blacklozenge 5$, we replaced the x that was after the $\blacklozenge$ with a 5? Well, let's do the same thing with the equation we're given for $\blacklozenge x$: Replace each x with a 5. We then get that $\blacklozenge 5 = 4(5) + 2 = 20 + 2 = 22$.

Let's try a practice problem. Look for what numbers we'll need to place in our original function, and where each number will go.

Question 10 of 20

If $x \sim y$ is defined as the sum of all of the prime integers between x and y, which of the following is equal to 17 ?

*Indicate **all** such values.*

☐ 1~10

☐ 4~10

☐ 6~12

☐ 14~18

Here's How to Crack It

Here we have a completely made-up mathematical term. This question says that whenever you see a ~, you must find the sum of all the prime integers between those two numbers. Rather than write out all the primes and figure out which numbers we could pick to get 17, let's PITA. As we plug in each answer choice, we're going to have to find the sum of all the prime integers between those two numbers. So 1~10 is the sum of all the prime integers between 1 and 10, which is $2 + 3 + 5 + 7 = 17$, so (A) is an answer. $4\sim10 = 5 + 7 = 12$, so cross off (B). $6\sim12 = 7 + 11 = 18$, so cross off (C). For the last answer, the only prime integer between 14 and 18 is 17, and since the sum of 17 and nothing is 17, (D) is also an answer. The answers are therefore (A) and (D).

You may also see function questions of the form $f(x) = x^2 + 5$. In that case, if they ask for $f(3)$, then wherever there used to be an x in our original function, we'll put a 3: $f(3) = 3^2 + 5 = 9 + 5 = 14$.

Whenever you see something unfamiliar on a GRE question, look to see if the question itself tells you what to do, and follow those directions.

FACTORIALS!

The term $n!$ is referred to as "n factorial," and whenever a factorial shows up on the GRE it pertains to the number of ways a number of elements can be chosen, or *arranged*. We'll discuss this further when we get to the section on permutations and combinations, a little later. But for now:

> The term $n!$ (read as "n factorial") represents the product of all integers from n to 1, inclusive. For example, $5! = 5 \times 4 \times 3 \times 2 \times 1$, or 120.

Factorial questions can look like they require more work than they really do. This is because you can usually cancel a lot of numbers and make a huge sequence of multiplications into something much more manageable. Here's an example:

Question 16 of 20

Quantity A	Quantity B
$\frac{24!}{23!}$	4!

Trigger: Factorials with division.

Response: Expand factorial and reduce.

- ○ Quantity A is greater.
- ○ Quantity B is greater.
- ○ The two quantities are equal.
- ○ The relationship cannot be determined from the information given.

Here's How to Crack It

When evaluating Quantity A, it might look like you're about to spend 10 minutes multiplying all those numbers on your calculator. Break the fraction down first, though, and you'll see that all but one of the numbers cancels out.

$$\frac{24!}{23!} = \frac{24 \times 23 \times 22 \times 21 \times \ldots \times 3 \times 2 \times 1}{23 \times 22 \times 21 \times \ldots \times 3 \times 2 \times 1} = 24$$

As you can see, you can cancel out everything from the 23 on, and you're left with 24. Quantity B also equals 24 (4 × 3 × 2 × 1), so the answer is (C).

Trigger: Factorials with addition or subtraction.

Response: Factor out common factorials.

Sometimes, however, a GRE question will involve adding or subtracting factorials. In that case, we're going to have to factor.

Question 10 of 20

What is the value of the expression $\dfrac{15!-14!}{13!}$?

[ ]

Here's How to Crack It

15! is too large to enter into the calculator, so we're going to have to factor out what 15! and 14! have in common. Since 15! is the same as 15 × 14 × 13 × 12 × 11 × 10... and 14! is 14 × 13 × 12 × 11... , we can rewrite 15! as 15(14!). Both 15(14!) and 14! contain 14!, so we can rewrite the fraction as $\dfrac{14!(15-1)}{13!}$. Note that if we distributed the 14! to each term within the parentheses, we'd have back our original numerator: 15(14!) – 14! We can simplify the 15 – 1 inside the parentheses, giving us $\dfrac{14!(14)}{13!}$. Now it's time to cancel out the two factorials, as we did with Question 16. $\dfrac{14(13!)(14)}{13!} = \dfrac{14(14)}{1} = 196$.

Factorials Quick Quiz

Question 1 of 4

Evaluate $\dfrac{25!}{23!} \times \dfrac{1}{4!}$

[ ]

Click on the answer box, then type in a number.
Backspace to erase.

Question 2 of 4

Each of the following is equivalent to 6! EXCEPT

- ○ $3! \times 5!$
- ○ $4! \times 30$
- ○ $6^2 \times 20$
- ○ $\frac{7!}{7}$
- ○ $\frac{12!}{2!}$

Question 3 of 4

Quantity A	Quantity B
$17! - 14!$	$4078 \times 14!$

- ○ Quantity A is greater.
- ○ Quantity B is greater.
- ○ The two quantities are equal.
- ○ The relationship cannot be determined from the information given.

Question 4 of 4

$\frac{(n+1)!}{(n-2)!} =$

- ○ 0
- ○ $n!$
- ○ $n^3 - n$
- ○ $n^3 - 4n$
- ○ $(2n)!$

Explanations for Factorials and Arrangements Quick Quiz

1. The first term, $\frac{25!}{23!}$, is equivalent to 25 × 24, and the second term equals $\frac{1}{24}$. Therefore, the final term is $\frac{25 \times 24}{24}$, or 25.

2. All of the answer choices are equivalent to 6!, or 720, except (E), because $\frac{12!}{2!}$ is the same as $\frac{12 \times 11 \times 10 \times 9 \times 8 \times 7 \times 6 \times 5 \times 4 \times 3 \times 2 \times 1}{2 \times 1}$. Because 12 × 11 × 10 is 1,320, it's possible to see that this is far bigger than 720 without even bothering with your calculator. The answer is (E).

3. Since there are no variables in this question, we can eliminate answer (D). We've got factorials combined with subtraction, which means we're going to have to factor. 17! is the same as 17 × 16 × 15 × 14!. We can therefore rewrite Quantity A as 17 × 16 × 15 × 14! – 14!. Since we have 14! in both terms, we can factor out 14!, giving us 14! (17 × 16 × 15 – 1). Now it's time for a bit of calculator work, which gives us 14!(4,080 – 1) = 14!(4,079). Both quantities have 14!, so let's focus on the other parts. Since 4,079 is larger than 4,078, 14!(4,079) is larger than 4,078 × 14!, and the answer is (A).

4. Rather than mess with the algebra here, just plug in a value of n (and make sure it's greater than 2). If $n = 5$, then $\frac{(n+1)!}{(n-2)!} = \frac{6!}{3!}$, or 120 (target answer). When you plug 5 into the answer choices, $n^3 - n = 5^3 - 5 = 125 - 5$, or 120. The answer is (C).

PROBABILITY

In probability questions, something that you *want* to happen is a "favorable outcome," and all of the things that *could* happen are "possible outcomes." So for these types of questions, your job is to figure out the chance that a favorable outcome—an outcome you want to happen—will occur.

Trigger: The word "probability."

Response: For each event, find the number of outcomes you want to happen, and divide by the total number of outcomes. $\frac{\text{want}}{\text{total}}$

The probability of a favorable outcome is found by dividing the number of possible favorable outcomes by the total number of possible outcomes.

- If a favorable outcome is impossible, then the probability that it will happen is 0.
- If a favorable outcome is a certainty, then the probability that it will happen is 1.
- The probability that something will happen plus the probability that it will not happen is equal to 1.
- If more than two outcomes are possible, the sum of the probabilities of all outcomes is equal to 1.

Question 19 of 20

Quantity A	Quantity B
The probability that the jack of diamonds will be selected at random from a standard deck of cards	The probability that Rhode Island will be selected at random from a list of U.S. states

- ○ Quantity A is greater.
- ○ Quantity B is greater.
- ○ The two quantities are equal.
- ○ The relationship cannot be determined from the information given.

Here's How to Crack It

There is only 1 jack of diamonds in a deck of 52 cards, so the probability of selecting it is $\frac{1}{52}$. Rhode Island is only 1 of 50 American states, so the probability that it will be selected at random is $\frac{1}{50}$. Because the denominator in Quantity A is greater, the fraction must be smaller. Therefore, the answer is (B).

Multiple Probabilities

If you're asked to find the probability that two specific events will occur one after the other, first find the individual probabilities that each event will occur, and then multiply them. There are two different scenarios on problems like these; sometimes the odds change, and sometimes they don't.

$$\textit{Probability of A and B} = \textit{Probability of A} \times \textit{Probability of B}$$

For example, if a problem involves flipping a coin, the probability that it will come up heads will always be $\frac{1}{2}$. It will never change, because there will always be heads on one side of the coin and tails on the other. Let's go through a typical problem you might see on the GRE.

Question 13 of 20

What is the chance of rolling a 5 on a standard, six-sided die three times in a row?

- ○ $\frac{1}{6}$
- ○ $\frac{1}{36}$
- ○ $\frac{1}{96}$
- ○ $\frac{1}{126}$
- ○ $\frac{1}{216}$

Here's How to Crack It

There is only one 5 on a standard, six-sided die, so the chance of rolling a 5 is $\frac{1}{6}$. When you roll the die a second or third time, you still have a $\frac{1}{6}$ chance of rolling a 5, so the chance remains $\frac{1}{6}$. Therefore, the chance of rolling three 5's in a row is $\frac{1}{6} \times \frac{1}{6} \times \frac{1}{6} = \frac{1}{216}$. The answer is (E).

Removing Items Changes the Probability

So far, our total, the denominator of our fraction, has always stayed constant. However, there will be some questions in which we remove items as we go. These problems often contain the phrase *without replacement,* because items are taken without replacing them. In that case, the total will change. At each point, we'll figure out the next probability by assuming that what the problem wants to happen has happened.

For instance, imagine there are 3 red marbles and 1 green marble in a bag. What's the probability of selecting 2 red marbles in a row? When we first reach into the bag, we'll have a $\frac{3}{4}$ chance of pulling out a red marble: There are 3 red marbles, out of a total of 4 marbles. But what about when we reach in again? We'll have to assume that we pulled out that first red marble. (Otherwise, who cares if we pull out a second red marble? We wanted two red marbles in a row, not a something else and then a red marble.) That means we have only 2 red marbles left, out of a total of 3 marbles, and our chance to pull that second red marble is $\frac{2}{3}$. Now we can combine our two probabilities by multiplying them together: $\frac{3}{4} \times \frac{2}{3} = \frac{1}{2}$.

Let's try a harder problem that uses a changing total.

Question 15 of 20

A shopping bag contains 5 green peppers and 4 red peppers. The grocer reaches in without looking and removes three of the peppers, one at a time.

Quantity A	Quantity B
The probability that the first two peppers he pulls out will be green and the third will be red	The probability that the first pepper he pulls out will be green and the last two will be red

- ○ Quantity A is greater.
- ○ Quantity B is greater.
- ○ The two quantities are equal.
- ○ The relationship cannot be determined from the information given.

Here's How to Crack It

This question is different because peppers are removed, so the odds change after each pepper leaves the bag. Work with Quantity A first: When the grocer reaches in for the first time, he has a $\frac{5}{9}$ chance of selecting a green pepper. The second time, there are only 4 green peppers out of a total of 8, so the chance of getting a second green pepper is $\frac{4}{8}$. The third time, there are only 7 peppers left and 4 of them are red. The chance of getting a red is $\frac{4}{7}$, and the composite probability is $\frac{5}{9} \times \frac{4}{8} \times \frac{4}{7}$.

Before you multiply, look at Quantity B. The successive probabilities are $\frac{5}{9}$ (green), $\frac{4}{8}$ (first red), and $\frac{3}{7}$ (second red), so the composite probability is $\frac{5}{9} \times \frac{4}{8} \times \frac{3}{7}$. When comparing the two quantities, the denominators are the same but the numerator in Quantity B is smaller. Therefore, Quantity B itself must be smaller, and the answer must be (A).

At Least Probabilities

Some questions will not ask for the probability of a single event, but instead ask for the probability that "at least" a certain number of events will happen. For instance, let's go back to the old, worn bag of marbles. Say there are 4 red marbles and 5 green marbles, and you're asked the likelihood of reaching in, taking out 3 marbles, and getting at least 1 red marble.

We *could* find that out by figuring out the probability of getting 1 red marble and then 2 green marbles, and adding that to the probability of getting a green, a red, and a green, plus the probability of getting a green, a green, and a red. But we could get more than just one red, right? We'd also have to find the chance of RRG, RGR, and GRR, and then the probability of getting RRR, and add all of those different ways of getting at least one red marble together.

But there's an easier way. Think of it this way: If there's a 30% chance of rain, what's the chance it won't rain? 70%. Since any two mutually exclusive probabilities always add up to 100%, we can sometimes find the probability that something *won't* happen rather than find the probability it will happen. Once we do, we'll subtract it from 1, because 100% = 1.

Trigger: Probability question asks "at least."

Response: Find the probability event *won't* happen, and subtract from 1.

Going back to our marble example, if you were asked "Did you pull at least 1 red marble from the bag?" and you answered "No," then what marbles did you pull? If you didn't pull at least 1 red marble, then you pulled all green marbles. The probability of pulling all green marbles is $\frac{5}{9}\times\frac{4}{8}\times\frac{3}{7}=\frac{5}{42}$. But we don't want to pull all green marbles; we want the opposite of that. Therefore the probability of getting at least one red marble is $1-\frac{5}{42}=\frac{37}{42}$.

Probability Quick Quiz

Question 1 of 5

Of the 50 U.S. states, 23 border a great body of salt water (the Atlantic Ocean, the Pacific Ocean, or the Gulf of Mexico).

Quantity A	**Quantity B**
The probability that a state selected at random has a saltwater border	The probability that a state selected at random does *not* have a saltwater border

- ○ Quantity A is greater.
- ○ Quantity B is greater.
- ○ The two quantities are equal.
- ○ The relationship cannot be determined from the information given.

Question 2 of 5

If the letters A, R, S, and T are arranged in a row, what is the probability that a proper English word will be formed?

- ○ $\frac{1}{12}$
- ○ $\frac{1}{4}$
- ○ $\frac{5}{24}$
- ○ $\frac{9}{24}$
- ○ $\frac{2}{3}$

Question 3 of 5

Fred has three dimes, two nickels, and four quarters in his pocket. If he reaches in and selects two coins, one at a time and without replacement, what is the probability that he will get two dimes?

- ○ $\frac{1}{12}$
- ○ $\frac{1}{6}$
- ○ $\frac{1}{4}$
- ○ $\frac{1}{3}$
- ○ $\frac{1}{2}$

Question 4 of 5

Harry the Horse throws two fair, six-sided dice.

Quantity A	**Quantity B**
The probability that he will roll numbers that sum to 8	The probability that he will roll numbers that sum to 7

- ○ Quantity A is greater.
- ○ Quantity B is greater.
- ○ The two quantities are equal.
- ○ The relationship cannot be determined from the information given.

Question 5 of 5

The probability of event J occuring is 0.63. The probability that both events J and K occur is 0.42. What is the probability that event K will occur?

Give your answer as a fraction.

Explanations for Probability Quick Quiz

1. There are 50 different states; these are the *possible* outcomes. There are 23 states with a saltwater border, and these are the desired outcomes. Therefore, the probability that a randomly selected state will have a saltwater border is $\frac{23}{50}$. If 23 have saltwater borders, then 27 states do not. This is the desired outcome for Quantity B. The number of possible outcomes remains the same, so the probability for Quantity B is $\frac{27}{50}$. This fraction is larger, so the answer is (B).

2. There are 24 (or 4!) ways to arrange the four letters, which can form 5 proper English words (arts, rats, star, tars, and tsar). The probability is $\frac{5}{24}$. The answer is (C).

3. There are 9 coins in Fred's pocket, 3 of which are dimes, so the probability that he will get a dime the first time is $\frac{3}{9}$. The second time, there are only 2 dimes out of 8 coins, so the probability drops to $\frac{2}{8}$. The composite probability is $\frac{3}{9}\times\frac{2}{8}$, or $\frac{6}{72}$. This reduces to $\frac{1}{12}$, so the answer is (A).

4. When two fair, six-sided dice are thrown, there are 36 possible rolls. Of these, 5 will sum to 8 (2 and 6, 3 and 5, 4 and 4, 5 and 3, and 6 and 2) and 6 will sum to 7 (1 and 6, 2 and 5, 3 and 4, 4 and 3, 5 and 2, and 6 and 1). Because there are more rolls of 7, the probability must be higher. The answer is (B).

5. Remember that the probability of two events happening is probability of A × probability of B. Since we know that probability of J × probability of $K = 0.42$, and we know the probability of J is 0.63, we know that 0.63 × probability of $K = 0.42$. Dividing both sides by 0.63 gives us that the probability of $K = \frac{0.42}{0.63} = \frac{42}{63} = \frac{2}{3}$.

As we near the homestretch, here are two more topics that could appear on the GRE. These topics occur very rarely, and when they do appear you'll see at most one question about them. It pays to reference them here, at the end of the chapter. Better safe than sorry, right?

GROUPS

There are two basic types of group questions: those that involve overlap between two groups, and those that involve groups in which there is no overlap. The group questions with overlap can be represented visually using the classic Venn diagram below, but you may find it easier, and more straightforward, to use an equation.

Group Formula:

Total = [Group 1] + [Group 2] – [Both] + [Neither]

Trigger: Group problem with overlap.

Response: Write down Group Formula.

The first thing you'll have to do with this sort of question is to recognize that you need to use the Group Formula. If there are any elements that overlap two of the groups, then we will need to use the Group Formula. We may not have any elements that are in neither group (Neither = 0), but if we have any overlapping elements, then those elements are being counted twice: Once because they belong to Group 1, and again because they belong to Group 2. To eliminate one of the times those Both elements are being counted, we subtract the Both. Note that Group 1 includes everything in Group 1, including those that are in Both. The same applies to Group 2.

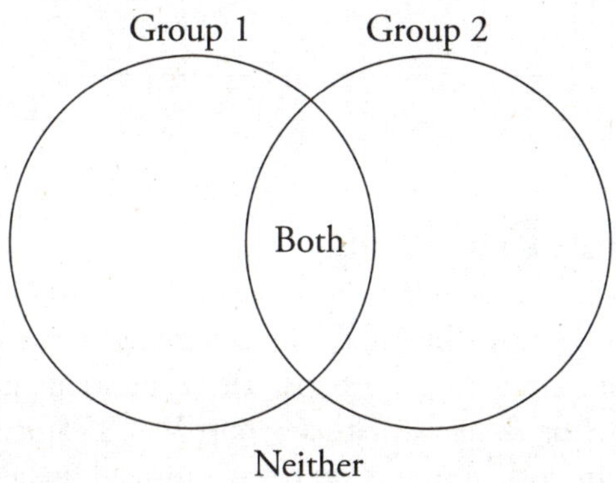

Let's try a problem using the Group Formula:

Question 8 of 20

Of the 213 people who responded to a genealogical poll, 82 said they have French ancestors and 113 said they have Italian ancestors. If 51 of the respondents said they have neither French nor Italian heritage, how many have both?

- ○ 33
- ○ 52
- ○ 69
- ○ 101
- ○ 124

Here's How to Crack It

Assign the groups (Group 1 is French, Group 2 is Italian), and plug the numbers you know into the formula: $82 + 113 - B + 51 = 213$. Because $246 - B = 213$, B must equal 33. The answer is (A).

Either/Or Group Problems

Trigger: Group question with no "Both Group *A* and *B*" elements.

Response: Draw Group Table.

The Group Formula is great, but it only works on problems in which there is some overlap between two groups. The GRE will also sometimes ask group questions in which there is no overlap between groups. These questions involve elements that are in either group *A* or *B*, *and* in either group *X* or *Y*. Since nothing is in both groups *A* and *B*, or in both groups *X* or *Y*, we'll instead use a simple chart, which looks like this:

	Group *X*	**Group *Y***	**Total**
Group *A*			
Group *B*			
Total			

These questions often involve elements that have two different qualities. For instance, cupcakes that are either vanilla or chocolate, and either frosted or unfrosted. Notice that there's no overlap within each group: A cupcake can't be both vanilla and chocolate, and it can't be both frosted and unfrosted. We could have vanilla frosted cupcakes, vanilla unfrosted cupcakes, chocolate frosted cupcakes, or chocolate unfrosted cupcakes.

Here's an either/or and either/or question example.

Question 10 of 20

Out of the 12,000 students and faculty at a certain university, there are 5 times as many students as there are faculty. There are 200 more female faculty members than there are male faculty members. If the university contains a combined total of 6,425 male students and faculty, how many female students are there?

Here's How to Crack It

Since you can be either a student or faculty or either male or female, and there's no overlap, we can draw our Group Table. It should look roughly like this:

	Male	Female	Total
Students			
Faculty			
Total			12,000

Now let's take the question apart in bite-sized pieces. The first thing we find out is that there are 5 times as many students as faculty. If we say there are f total members of the faculty, then $5f = s$. Now that we know there's a total of 12,000 faculty and students, $12{,}000 = f + s$. Substituting in our earlier equation, we get $12{,}000 = f + 5f$. Since $12{,}000 = 6f$, $f = 2{,}000$. Since there are 2,000 faculty members, there are 10,000 students.

	Male	Female	Total
Students			10,000
Faculty			2,000
Total			12,000

Now we can use the next piece of information: There are 200 more female faculty members than there are male faculty members. There are 2,000 total, so $m + w = 2{,}000$. We know that $w = 200 + m$, which we can plug into our earlier equation to get $m + 200 + m = 2{,}000$, so $m = 900$. If there are 900 male faculty, there are (2,000 – 900) 1,100 female faculty.

	Male	**Female**	**Total**
Students			10,000
Faculty	900	1,100	2,000
Total			12,000

Okay, we've got one more piece to use: There's a total of 6,425 males. There's a total of 12,000 students and faculty, which means there are (12,000 – 6,425) 5,575 females. Filling that into our chart, we get:

	Male	**Female**	**Total**
Students			10,000
Faculty	900	1,100	2,000
Total	6,425	5,575	12,000

Now we can easily fill in the rest of the table with simple subtraction: The 6,450 – 900 = 5,550 males, and the number of female students is (5,575 – 1,100) 4,475. The answer to the question is 4,475. (You don't need to know the number of male students to answer the question, but for the sake of completeness, 6,425 – 900 = 5,525 male students.)

	Male	**Female**	**Total**
Students	5,525	4,475	10,000
Faculty	900	1,100	2,000
Total	6,425	5,575	12,000

STANDARD DEVIATION

Standard deviation means almost exactly what it looks like it means: deviation from the "standard," or mean, value of a set of numbers. A "normal" distribution of data means that most of the numbers in the data are close to the mean and few "outliers" skew toward the extremes. The bigger the deviation from the norm, the wider the spectrum of numbers involved.

The Bell Curve

A "normal" distribution is best displayed in the form of a regular bell curve, which looks like this:

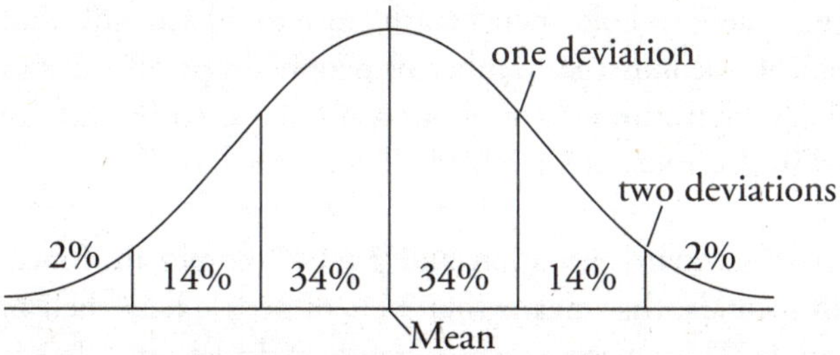

The mean is the middle number, right at the 50% mark. The GRE will either just present you with the mean straight up, or it will be one quick calculation away. The rest of the lines on the curve represent standard breakpoints at 34%, 14%, and 2%. These mean that, within a normal distribution, 68% of the group (34% on the left, 34% on the right) are within one standard deviation from the mean.

Trigger: Question contains the words "normal distribution" or "standard deviation."

Response: Draw a bell curve and label the mean and the 34-14-2 points for each standard deviation.

To explain the normal distribution, let's use a sample data set. Say we asked 1,000 people to see how long they could hold their breath. After measuring all those people, and watching all those faces turn purple, we calculated that the average (arithmetic mean) number of seconds that people could hold their breath was 50 seconds. We then handed our data to a statistician friend, who calculated that our data followed a normal distribution and the standard deviation was 15 seconds.

Since the mean time was 50 seconds, half of the people were able to hold their breath for less than 50 seconds and half the people were able to hold their breath for longer. Our data looks something like this:

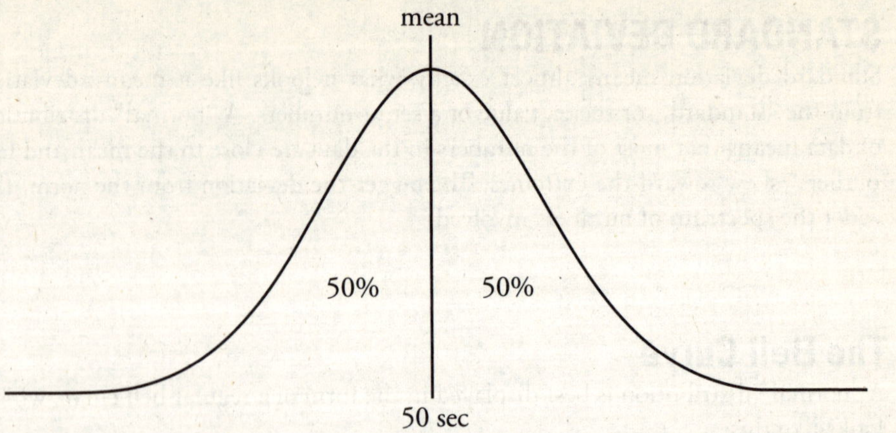

Most people were able to hold their breath close to 50 seconds, and as we get farther away from 50 seconds the number of people drops off substantially in both directions. That's what's important about a normal distribution. The data points are distributed in this nice, predictable bell curve shape.

Here's where our standard deviation and 34-14-2 comes in. Since our standard deviation is 15 seconds, that means that 34% of people held their breath between 50 and 65 seconds. Those people were 1 standard deviation or less from the mean. When we move another standard deviation (15 seconds) away, we find that only 14% of people could hold their breath for anywhere between 65 and 80 seconds. Only 2% of people (all of whom have excellent lungs and do plenty of cardio) could hold their breath for more than 80 seconds.

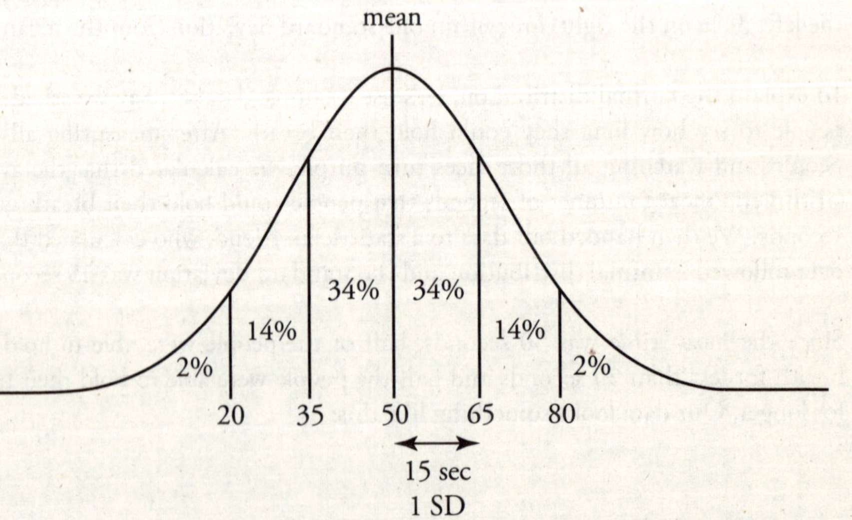

The same holds true in the other direction. 34% of people surveyed were able to hold their breath for anywhere between 35 and 50 seconds. Those people were 1 standard deviation or less from the mean. As we move another standard deviation of 15 seconds away, we find that only 14% of people could hold their breath for between 20 and 35 seconds. Finally, we have the bottom 2% of people. These people are 2 standard deviations or more below the mean, which is another way of saying that they are the absolute worst at holding their breath.

You will never have to calculate the standard deviation on the GRE, but you will have to understand how it works.

- **The bigger the standard deviation, the more spread apart the numbers.** Imagine an unkept field. If we went out and measured the height of the grass in that field, we'd probably get a standard deviation of about 20 centimeters or so: We'd have some really tall grass, but also some young, short grass. Because of those variations in height, we'd have a large standard deviation. The height of grass wouldn't always be too close to the mean. Now imagine we took a lawnmower to that field. The lawnmower wouldn't cut every blade of grass to the exact same height, but it would make every blade fairly close to the same height. Our standard deviation would shrink from 20 cm to 1 cm. Sure, some grass is a little taller than our mean height and some is a little shorter, but overall the heights of the blades of grass are very close to being the same. There aren't many variations in height, so we have a small standard deviation. Lots of tall grass, short grass, and medium grass meant a large standard deviation, but when all of our grass was close to the same height we had a small standard deviation.
- **The more standard deviations away from the mean, the stranger you are.** Say that the average pop song is three minutes long, with a standard deviation of 30 seconds. A song that is 3:30 isn't that weird, because it's only 1 standard deviation away from the mean. A song that's 3:45 is a little more unusual, because it's 1.5 standard deviations from the mean. A song that's 4:30 long is really weird, because that's 3 standard deviations away from the mean. For a pop song, it's really unusual to be that long. Compare that to the opera. The average opera is 90 minutes long, with a standard deviation of 25 minutes. A two-hour (120 minute) long opera, therefore, isn't that weird: It's a little more than 1 standard deviation away from the mean. But how about a five-minute opera? That's around 3.5 standard deviations below the mean. For an opera, that's freakishly short.

That's all you need to know about standard deviation. Sometimes, in fact, the GRE will even give you 34-14-2 and a drawing of the bell curve, but you should still memorize 34-14-2 and remember that the bigger the standard deviation, the more spread apart the numbers, and that the more standard deviations away from the mean, the stranger you are.

Question 16 of 20

The average age among members of a retirement community is 61, and the standard deviation is 2.5 years. Under a normal distribution, what percent of the community members are younger than 56 years old?

- ○ 2%
- ○ 16%
- ○ 32%
- ○ 34%
- ○ 50%

Here's How to Crack It

If the average age is 61 and the standard deviation is 2.5 years, then a person who is 56 years old is 2 deviations from the mean (2×2.5 is 5, and $61 - 5 = 56$). This means that only the bottom 2% of the group are less than 56 years old. The answer is (A).

Your scratch paper should look like this:

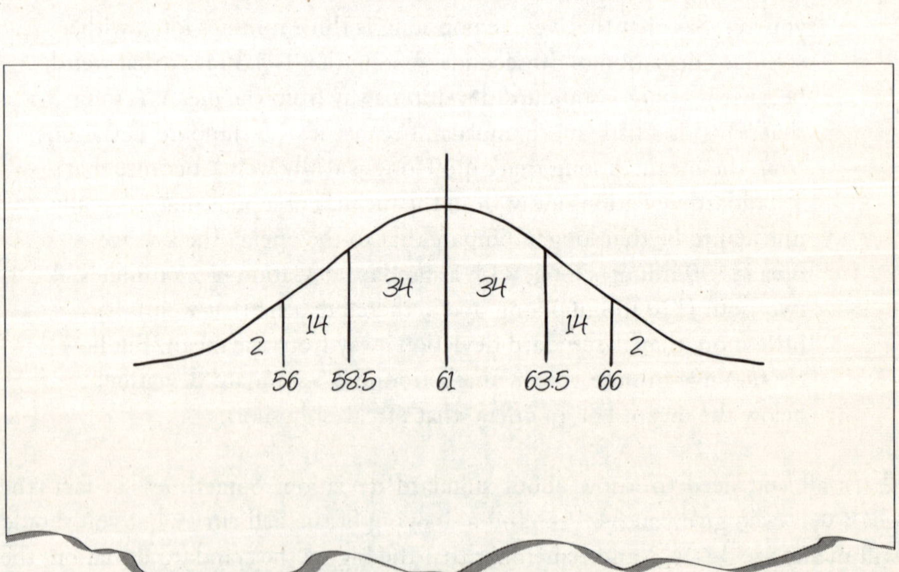

PERMUTATIONS AND COMBINATIONS

We've got one last topic to cover. It doesn't come up often, so if you're not completely comfortable with plugging in, PITA, and all those other topics, you should go back and review those first.

Good? Okay, so let's talk about permutations and combinations. Permutations and combinations questions typically ask about the number of different ways of either arranging things or grouping things. For these questions, we'll pretend that we're arranging or grouping our items piece by piece, item by item, element by element.

Each time we have to choose something, we'll figure out how many options we have. Then we'll pretend we chose one, and now have to choose the next thing. Once we've figured out how many choices we had at each point, we'll multiply those numbers together.

Trigger: The phrases "arrangements," "permutations," "combinations," "different ways," "many ways," or "different groups" appear in the problem.

Response: Draw a horizontal line for each choice we have to make.

Let's do an easy example first. Pretend you have three pictures, one of Groucho, one of Harpo, and one of Chico, that you need to place on a shelf. To find the number of different ways to arrange those pictures, you could simply write out all the different possibilities. We could have Groucho, Harpo, and then Chico, or we could have Groucho, Chico, then Harpo. We could also have Harpo, Groucho, Chico, or Harpo, Chico, Groucho, or Chico, Harpo, Groucho, or Chico, Groucho, Harpo. That's six different ways of arranging those pictures.

So let's look at how we listed our original set of three pictures, and do some math. We have three different choices we have to make: Who is going in the first spot on the wall, who is going in the second, and who is going into the third. Because we have three choices to make, we'll draw three slots, short horizontal lines to put numbers in, on our scratch paper, like so:

_____ _____ _____

The first thing we had to decide was whether to put Groucho, Harpo, or Chico first. We had three different options of whom to place in that first spot, so we'll put three in the first slot on our scratch paper:

__3__ _____ _____

Notice that once we choose Groucho to go first, we have only two options left: Does Harpo go next, or does Chico? The same is true if we put Harpo or Chico first; we have only two options for that second slot. Put a 2 on that second spot.

__3__ __2__ _____

Once we've put Groucho in the first spot, and Harpo in the second, we've got to put Chico last, right? Or, say we put Harpo in the first spot and Chico in the second: In that case, Groucho's got to be last. Since, no matter whom we place in the first two spots, we've only got one option (whoever is left) for the last spot, put a 1 in the last slot.

$\underline{\quad 3 \quad}$ $\underline{\quad 2 \quad}$ $\underline{\quad 1 \quad}$

Now multiply those numbers together. $3 \times 2 \times 1 = 6$. Six different ways, exactly how we calculated when we listed everything out earlier.

Each time we had a choice to make, we looked at how many options we had for that choice. As we moved through our choices, we had fewer options, because we pretended that the previous choices had been made.

Try the next problem.

Question 8 of 20

Eight horses compete in a race. The first horse to finish wins a blue ribbon, the second wins a red ribbon, and the third wins a yellow ribbon. In how many ways can the ribbons be distributed?

[ ]

Here's How to Crack It

The end of the problem states *in how many ways*, which means this is an arrangement question. We've got 3 ribbons to give away, so draw 3 slots on your scratch paper. For that blue ribbon, we could give it to whichever of those 8 horses comes in first, which means we've got 8 options for our first slot. Once we've given one of the horses a blue ribbon, there are only 7 horses left to win the red ribbon, so put a 7 in the second slot. Now there are 6 horses left to win the yellow ribbon, so put 6 in the third slot. We now have $8 \times 7 \times 6 = 336$ different ways of distributing those ribbons.

If a problem contains restrictions on what we can choose at certain points, we'll start with those restricted positions first, and then deal with everything else. It's as though you were seating guests around a table. If you knew that two people just really, really, hated each other, then the first thing you'd do is make sure they sat on opposite ends of the table. Once that's taken care of, you'd then place the rest of your friends, the ones who do not hold grudges over trivial events that no one really remembers, in the remaining seats.

Question 13 of 20

Boone is making a list of songs to listen to. He has 5 songs, but only 2 of the songs are suitable for the last song of the list. How many different lists could Boone make of his 5 songs?

- ○ 24
- ○ 48
- ○ 60
- ○ 120
- ○ 3,125

Here's How to Crack It

The problem asks for *how many different lists*, so this is another arrangement question. We've got 5 songs we need to put in order, so draw 5 slots on your scratch paper. Here, however, we're limited into what can go in the last spot in our list: It's got to be one of 2 possible songs. So we'll start by putting 2 in the last slot.

___ ___ ___ ___ 2

Now we can deal with the other slots. Since we have no other restrictions, let's continue by going back to the first slot. Since we put one of our songs in that last spot, we have only 4 songs left to put in the first slot:

4 ___ ___ ___ 2

The second slot therefore has only 3 songs possible, because we've already placed songs in the first and last slots. Continuing to fill out the slots, we get:

4 3 2 1 2

Multiplying $4 \times 3 \times 2 \times 1 \times 2$ gives us 48, answer (B).

The questions we've done so far are all *permutation* questions, because the order in which our elements are placed matters. Putting our Chico picture before our Harpo picture would be different from putting Harpo before Chico. Switching the first and second place ribbons of two of the horses would mean a different outcome of the race. Switching the middle songs of Boone's playlist would mean a different play-list than we had originally. Permutation questions are questions in which order matters.

What about when order doesn't matter? Those are called combination questions, and they often contain words such as "groups," "teams," or (obviously), "combinations." If you had to choose two books to loan, at the same time, to a friend, it wouldn't matter if you loaned your friend *Pale Fire* and *The Tin Drum* as opposed to loaning her *The Tin Drum* and *Pale Fire*. Either way, she's borrowing the same two books, and she better keep care of them.

Trigger: The words "team," "groups," "combinations," or order doesn't matter.

Response: At the end of the problem, divide the answer by the factorial of the number of slots.

Because order doesn't matter on combination questions, we're going to add one extra step at the end. We need to get rid of the doubles or triples from our groups, so we'll divide our answer by the factorial of our number of slots. An easier way to think of it is that we'll count down to 1 underneath our slots.

Question 2 of 20

A group of 4 is to be chosen from among 9 total employees.

Quantity A	Quantity B
The total number of different groups that could be chosen	120

○ Quantity A is greater.

○ Quantity B is greater.

○ The two quantities are equal.

○ The relationship cannot be determined from the information given.

Here's How to Crack It

Since we have a group of 4, we'll draw 4 slots on our paper. We could choose one of the 9 people for the first slot, leaving 8 for the second slot, 7 for the third, and 6 for the fourth, giving us $\underline{9} \times \underline{8} \times \underline{7} \times \underline{6}$. However, we're choosing *groups* of people, which means that the order we choose each group doesn't matter. As long as it's got the same people in it, who cares what order those people were chosen in?

Our last step, therefore, is to divide by 4! We'll show this by counting down from 4 underneath each slot, making each number a fraction. We now have $\frac{9}{4}\times\frac{8}{3}\times\frac{7}{2}\times\frac{6}{1}$, which we can simplify to $\frac{3}{1}\times\frac{2}{1}\times\frac{7}{1}\times\frac{3}{1}=126$. Since Quantity A is larger than Quantity B, the answer is (A).

By the way, did you notice how the denominator of our fraction canceled out completely? That will always happen with combinations. Think of it this way: There will never be a fractional number of ways to select groups of people. If the denominators of the fractions don't all cancel out to 1, you may have either made an arithmetic mistake, or missed a chance to cancel out.

To review, if order matters ("arrangements," "schedules," et cetera), then it's a permutation. Draw a slot for each time you need to make a choice, and then fill in each slot with the total number of options for each choice.

If order doesn't matter ("groups," "pairs," "teams," et cetera), then it's a combination. Start the problem as you would a normal permutation problem, but then divide by the factorial of the number of slots you have.

Permutations and Combinations Quick Quiz

Question 1 of 5

At a certain ice cream shop, a sundae contains two scoops of different varieties of ice cream. If the shop sells 20 different varieties of ice cream, how many different sundaes are possible?

- ○ 40
- ○ 190
- ○ 200
- ○ 380
- ○ 400

Question 2 of 5

A delivery driver must make 8 stops over the course of the night.

Quantity A	Quantity B
The number of different routes connecting all 8 stops that the delivery driver could take	40,320

- ○ Quantity A is greater.
- ○ Quantity B is greater.
- ○ The two quantities are equal.
- ○ The relationship cannot be determined from the information given.

Question 3 of 5

A website requires a password made up of 2 one-digit numbers followed by 3 of the 26 letters. How many different passwords are possible if repetition of numbers and letters is allowed?

- ○ 98
- ○ 17,576
- ○ 117,000
- ○ 1,404,000
- ○ 1,757,600

Question 4 of 5

A chef has lamb, carrots, potatoes, celery, and peppers. From these, she chooses three different ingredients for a stew. She will serve the stew with bread, salad, or dumplings. How many total different meals could the chef serve?

[ ]

Question 5 of 5

Ten runners compete in a race. The first 6 to finish receive a $100 prize. In addition, the first 3 to finish receive medals: gold for first, silver for second, and bronze for third.

Quantity A	Quantity B
The number of ways of distributing the $100 prize	The number of ways of distributing the three medals

- ○ Quantity A is greater.
- ○ Quantity B is greater.
- ○ The two quantities are equal.
- ○ The relationship cannot be determined from the information given.

Explanations for Permutations and Combinations Quick Quiz

1. Since each sundae contains two scoops of ice cream, draw two slots on your scratch paper: __ __. There are a total of 20 different flavors, so we have 20 options for our first slot: 20 __. The problem states that the sundae must have two different types of ice cream, so we can't have the same flavor in the second spot as in the first. Therefore, there are only 19 different options left for our second choice: 20 19. Before we multiply, we have to ask ourselves if order matters. Would a sundae with a scoop of chocolate and a scoop of vanilla be different than a sundae with a scoop of vanilla and a scoop of chocolate? Nope. Order doesn't matter, so this is a combination. Therefore, we're going to have to count down from two underneath our slots, giving us $\frac{20}{2} \times \frac{19}{1} = \frac{10}{1} \times \frac{19}{1} = 190$, answer (B).

2. As the driver decides his route, he's going to have 8 choices to make: Which stop to make first, which to make second, and so on. Draw 8 slots on your scratch paper: __ __ __ __ __ __ __ __. For his first stop, he has 8 options of where to stop, so put an 8 in the first slot. Once he's made that stop, he has only 7 options left for the second stop, so put 7 in the second slot. Continuing to fill in the slots, we have $\underline{8}$ $\underline{7}$ $\underline{6}$ $\underline{5}$ $\underline{4}$ $\underline{3}$ $\underline{2}$ $\underline{1}$. Does order matter? Definitely. If the driver stops at House A before House B, that's a different route than if he stops at House B before House A. This a permutation. Since order matters, we can multiply together $\underline{8} \times \underline{7} \times \underline{6} \times \underline{5} \times \underline{4} \times \underline{3} \times \underline{2} \times \underline{1}$, using the calculator and scratch paper, giving us 40,320. Since the quantities are equal, the answer is (C).

3. The password has 5 total numbers or letters, so we'll have to make 5 choices. Draw 5 slots on your scratch paper: __ __ __ __ __. The first thing we'll have to choose is a one-digit number. Since there are 10 one-digit numbers, from 0 to 9, we have 10 options for our first slot. We also need to place a one-digit number in the second spot. However, since the problem states that repetition is allowed, we have 10 options for our second slot as well. In other words, if we choose 4 as the first character of our password, we're allowed to choose any one-digit number, including 4, as the second character. So far, we've got $\underline{10}$ $\underline{10}$ __ __ __. Now we get into the alphabet. Our first letter could be any of the 26 letters, so put 26 in the third slot. The second letter could also be any of the 26 letters, since repetition is allowed, so put 26 in the fourth slot. The last character of our password also needs to be a letter, with repetition allowed, so put 26 in the last spot. Our slots are now $\underline{10}$ $\underline{10}$ $\underline{26}$ $\underline{26}$ $\underline{26}$. Does order matter? Having 12ABC as a password is different than having 21CBA, so order definitely matters. Multiply together the numbers in our slots: $\underline{10} \times \underline{10} \times \underline{26} \times \underline{26} \times \underline{26} = 1{,}757{,}600$, answer choice (E).

4. This is a slightly more difficult problem, because we've got two overall decisions to make: which stew to serve and what to serve it with. But just choosing the stew itself will take some math. Let's start with the stew. The chef has 3 options of what to put in the stew, so we have 3 slots: __ __ __. There are 5 ingredients, so we have 5 options for the first slot. Since we must use 3 *different* ingredients, we have only 4 options left for the second slot, 3 options for the third slot, and our slots are $\underline{5}$ $\underline{4}$ $\underline{3}$. Does order matter? Putting lamb, celery, and potatoes in the stew is the same as putting potatoes, celery, and lamb, so order doesn't matter, so this portion of the problem is a combination. Therefore, we have to count down from 3 underneath our slots, giving us $\frac{5}{3} \times \frac{4}{2} \times \frac{3}{1} = \frac{5}{1} \times \frac{2}{1} \times \frac{1}{1} = 10$ different stews. Now we have to decide our different meals. We'll have to choose a stew and a side, so we have two slots: __ __. For our first slot, we have the option of

any of those 10 delicious stews we calculated earlier. For our second slot, we can choose 1 of those 3 sides. The slots are $\underline{10}$ $\underline{3}$. Does order matter? Definitely, because we've got two different types of items here: We can't serve dumplings as a stew, and we can't serve a stew in place of bread. $\underline{10} \times \underline{3} = 30$ different meals.

5. Let's focus on Quantity A first. We're giving away 6 prizes, so we have 6 slots: __ __ __ __ __ __. For the first slot, we can give the prize to any of the 10 runners in the race. Once we've given away that prize, we only have 9 runners left who can receive a prize, and so on, giving us $\underline{10}$ $\underline{9}$ $\underline{8}$ $\underline{7}$ $\underline{6}$ $\underline{5}$. Does order matter? Well, if I come in first and get \$100, is that any different than coming in sixth and getting \$100? Not really, since I'm getting \$100 either way. Since order doesn't matter, this is a combination question, and we'll have to count down from 6 underneath our slots: $\frac{10}{6} \times \frac{9}{5} \times \frac{8}{4} \times \frac{7}{3} \times \frac{6}{2} \times \frac{5}{1}$. Before you multiply, see which denominators cancel out with which numerators, and we have $\frac{2}{1} \times \frac{3}{1} \times \frac{1}{1} \times \frac{7}{1} \times \frac{1}{1} \times \frac{5}{1} = 210$ different ways of handing out the six \$100 prizes.
Now let's work on Quantity B. We have 3 medals to give away, so we've got 3 slots: __ __ __. The gold medal could go to any of the 10 runners, so put 10 in the first slot. The silver medal can go to any of the remaining 9 runners, and the bronze can go to any of the remaining 8 runners, so the slots are $\underline{10}$ $\underline{9}$ $\underline{8}$. Does order matter? If two runners, Dave and Rob, compete and Dave wins the gold medal and Rob wins the bronze, is that the same as Rob winning the gold and Dave winning the bronze? Nope, so order matters. This is a permutation and we have to multiply together the slots: $\underline{10} \times \underline{9} \times \underline{8} = 720$. Quantity B is larger than Quantity A, so the answer is (B).

Congratulations! You've made it through the most arcane topics that the GRE tests. If you can handle yourself here, you're in great shape for the quantitative section. If your math score is average, then it's highly unlikely that you'll have to worry about much of the material here. If you usually score well, however, knowing this stuff could mean the difference between a good score and a great one.

Be sure to keep practicing; you're almost ready to take the test and get on with your life!

The Rest of the Story Drill

Question 1 of 15

All but 4 of the counselors on staff at a certain summer camp have sailing certification, first aid certification, or both. Twice as many counselors have neither certification as have both certifications, and 7 counselors have sailing certification. If there are a total of 22 counselors on staff, then how many of the counselors have first aid certification?

[ ]

Question 2 of 15

What is the value of $\frac{8!}{10!}$?

- $\frac{4!}{5!}$
- $\frac{1}{2!}$
- $9!$
- $\frac{1}{5(6!)}$
- $\frac{1}{15(3!)}$

Question 3 of 15

If set $X = \{12, 16, 20\}$, then which of the following sets has a standard deviation larger than that of set X ?

- {2, 4, 6}
- {4, 5, 6}
- {13, 16, 19}
- {20, 32, 44}
- {40, 40, 40}

Question 4 of 15

Ryan's bakery is selling 212 cakes. 131 cakes are chocolate and the rest are vanilla. 104 cakes have mocha frosting, and the rest have coconut frosting. If 37 of the chocolate cakes have mocha frosting, how many of the vanilla cakes have coconut frosting?

- 14
- 67
- 81
- 104
- 108

Question 5 of 15

If a value for the integer x is randomly selected and $-10 < x < 10$, what is the probability that x is even?

- $\frac{1}{19}$
- $\frac{1}{4}$
- $\frac{1}{2}$
- $\frac{9}{19}$
- $\frac{11}{21}$

Question 6 of 15

Which of the following is the least value of x for which $\frac{x!}{6!}$ is an integer greater than 1 ?

- ○ 2
- ○ 3
- ○ 6
- ○ 7
- ○ 12

Question 7 of 15

The purchaser of a certain car must choose 2 of 5 special options and 5 of 6 interior features. How many different combinations of options and features are available to the purchaser?

- ○ 10
- ○ 16
- ○ 18
- ○ 30
- ○ 60

Question 8 of 15

Laura has decided to display some of her glass animal figurines on a shelf that can hold 5 of them. If she has 6 circus animals and 5 farm animals to choose from, and she wants a farm animal in the middle of the display, then how many distinct arrangements of the figurines are possible?

- ○ 30
- ○ 150
- ○ 3,024
- ○ 25,200
- ○ 30,240

Question 9 of 15

Two six-sided dice with sides numbered 1 through 6 are rolled. If the two resulting numbers are multiplied, what is the probability that their product will be even?

- ○ $\frac{1}{12}$
- ○ $\frac{1}{4}$
- ○ $\frac{1}{2}$
- ○ $\frac{3}{4}$
- ○ $\frac{11}{12}$

Question 10 of 15

Sandy is designing an internet banner advertisement and has decided to use one background, one font, two different accent images, and four different colors in her design. If Sandy has 5 backgrounds, 4 fonts, 6 accent images, and 12 colors to choose from, then how many different advertisement designs can she make?

[ ]

Question 11 of 15

An Outdoor Adventure Camp offers 1-week sessions. During the first week, 44 campers go fishing, 33 go orienteering, and 37 do neither activity. The same number of campers goes orienteering in week 2 as in week 1, and 15 campers in week 2 do both activities. If 49 of the 120 campers who attend week 2 do neither activity, and twice as many campers attend week 1 as go fishing in week 2, then how many total campers attended these 2 weeks?

- ○ 196
- ○ 226
- ○ 234
- ○ 360
- ○ 662

Question 12 of 15

During this year's fundraiser, students who sell at least 75 subscriptions will win a prize. The fourth-grade students sold an average of 47 magazine subscriptions per student, and the sales have a standard deviation of 14. If the sales of subscriptions are normally distributed, then what percent of the fourth-grade students will receive a prize?

- ○ 0.02
- ○ 0.25
- ○ 2
- ○ 25
- ○ It cannot be determined from the information given

Question 13 of 15

A college admissions committee must select candidates for a certain program from among high school applicants and transfer applicants. The committee has already chosen 132 female candidates in a 3 to 1 ratio of high school students to transfer students, and will maintain this ratio in the selection of male candidates. If the final class size must be between 325 and 350 students, then which of the following is an acceptable number of male high school applicants the committee could choose?

*Indicate **all** such statements.*

- ☐ 99
- ☐ 144
- ☐ 156
- ☐ 162
- ☐ 168
- ☐ 180

Question 14 of 15

Robin and Terry want to invite 5 of their friends to their wedding. Robin has 7 friends, Terry has 6, and Robin and Terry have no friends in common. If at least 1 of Robin's friends and at least 1 of Terry's friends must be invited, how many different groups of friends could Robin and Terry invite to their wedding?

- ○ 462
- ○ 924
- ○ 1,260
- ○ 2,520
- ○ 151,200

Question 15 of 15

Tony's political science final exam consists exclusively of 8 true/false questions. If Tony guesses on every question, what is the probability that he gets exactly 7 questions right?

- ○ $\frac{1}{32}$
- ○ $\frac{1}{16}$
- ○ $\frac{1}{8}$
- ○ $\frac{7}{8}$
- ○ $\frac{31}{32}$

EXPLANATIONS FOR THE REST OF THE STORY DRILL

1. 13

First, write down the group formula, $Total = Group_1 + Group_2 + Neither - Both$, and fill in what you know. If sailing certification is $Group_1$ and first aid certification is $Group_2$, then $22 = 7 + Group_2 + 4 - 2$. So, $22 = Group_2 + 9$, and you can solve for $Group_2 = 13$.

2. E

$8! = 8 \times 7 \times 6 \times 5 \times 4 \times 3 \times 2 \times 1$, and $10! = 10 \times 9 \times 8 \times 7 \times 6 \times 5 \times 4 \times 3 \times 2 \times 1$. When dividing $\frac{8!}{10!}$, everything but 10×9 in the denominator cancels out, and you have $\frac{1}{10 \times 9} = \frac{1}{90}$. For choice (A), $4 \times 3 \times 2 \times 1$ cancels out of the top and bottom, leaving $\frac{1}{5}$. Choice (B) is $\frac{1}{2 \times 1} = \frac{1}{2}$. Estimate that choice (C) is greater than 1, and you need an answer less than 1. Choice (D) is $\frac{1}{5 \times 6 \times 5 \times 4 \times 3 \times 2 \times 1}$. Estimate that the denominator is greater than 90. Choice (E) is $\frac{1}{15 \times 3 \times 2 \times 1} = \frac{1}{90}$.

3. D

Don't calculate anything on this one. Instead, just remember that standard deviation is a measure of how much numbers in a set vary, or *deviate*, from the average, or *standard*. Since all of the sets contain three equally spaced numbers, the middle number in each case is the average. The other two numbers in the correct answer, choice (D), differ most from their average. The only comparably large numbers are in choice (E) and don't differ at all, and thus have a standard deviation of 0.

4. A

Take this problem one step at a time, using a grid layout if that helps. 131 cakes are chocolate, so 81 must be vanilla. 104 have mocha frosting, so 108 have coconut frosting. 37 chocolate cakes have mocha frosting, so there are 94 chocolate cakes with coconut frosting, leaving 14 vanilla cakes with coconut frosting.

5. D

There are 19 numbers in the range between –10 and 10. Of those, 9 are even. The probability of selecting an even number is $\frac{9}{19}$.

6. D

It's an algebra question with numbers for answer choices, so set up your scratch paper to Plug In the Answers. Start with choice (C): if $x = 6$, then $\frac{6!}{6!} = \frac{6 \times 5 \times 4 \times 3 \times 2 \times 1}{6 \times 5 \times 4 \times 3 \times 2 \times 1} = 1$; choice (C) is too small, so eliminate choices (A), (B), and (C). Now do the same thing for $x = 7$; all the numbers will cancel out of the denominator, and you'll be left with 7 in the numerator, so the whole expression equals 7. Choice (D) is correct. Choice (E) yields an integer as well, but isn't the *least value* among the choices to do so.

7. E

The purchaser has 5 options to choose from for his first choice of special options and 4 left to choose from for his second choice. $5 \times 4 = 20$ arrangements, but because the order of choice doesn't matter, divide that 20 by 2 to get 10 special option groups. Do the same thing for the interior features: $6 \times 5 \times 4 \times 3 \times 2 = 720$. Divide 720 by $(5 \times 4 \times 3 \times 2)$ because the order of choice doesn't matter. Multiplying the 10 special options by the 6 interior features yields 60 possible options. You could also use the combination formula, $C = \frac{n!}{r!(n-r)!}$, where n is the total number in the group you're drawing from and r is the number of those items you're selecting to find the combinations for special options and interior features.

8. D

There are 5 spaces to fill with a restriction on the middle space, so start with the restricted space. The middle space must be a farm animal, and there are 5 potential figurines for that spot, so 5 goes in that space. The remaining spaces are unrestricted, so there are 10 figurines (6 circus and 4 remaining farm animals) left from which to choose. Slot your first of four remaining spaces with 10, then 9, 8, 7 for the remaining spaces as each one fills. $10 \times 9 \times 5 \times 8 \times 7 = 25{,}200$, making choice (D) the best answer. Choice (E) is a trap answer if you don't account for the restricted space.

9. D

The only outcome that would not result in an even product is two odd numbers, because even × (even or odd) = even. Subtracting the probability of two odd rolls from 1 will give the probability that the product is even because P(odd product) + P(even product) = 1. The probability that a roll will be odd is $\frac{3}{6}$, or $\frac{1}{2}$. The probability that both will be odd is $\frac{1}{2} \times \frac{1}{2} = \frac{1}{4}$. The probability that the product will be even is $1 - \frac{1}{4} = \frac{3}{4}$.

10. 148,500

This is a tricky combination question because you have to treat each category as its own combination, and then multiply the results. Find the number of ways you can choose 1 out of 5 backgrounds, 1 out of 4 fonts, 2 out of 6 images, and 4 out of 12 colors, and then multiply those results together. So, $\frac{5}{1} \times \frac{4}{1} \times \frac{6}{2} \times \frac{5}{1} \times \frac{12}{4} \times \frac{11}{3} \times \frac{10}{2} \times \frac{9}{1}$ = 148,500.

11. B

The question mentions *neither* and *both*, so be sure to write out the group formula, *Total* = $Group_1$ + $Group_2$ + *Neither* – *Both*, for each week. Let fishing be $Group_1$ and orienteering be $Group_2$, and fill in what you know: for week 1, you have *Total* = 44 + 33 + 37 – *Both*; for week 2, you have 120 = $Group_1$ + 33 + 49 – 15. Solve for $Group_1$ in the latter equation, and 53 people went fishing in week 2. Twice that many, or 106, attended week 1 overall, and over the 2 weeks the camp was attended by a total of 106 + 120 = 226 people.

12. C

When you see the words *standard deviation* or *normally distributed*, draw your bell curve and fill in the percentages: 34, 14, and 2. The average of 47 and a standard deviation of 14 means 75 is 2 standard deviations above the mean, so 2 percent of the students will receive a prize.

13. C and D

Since you have two traits (female/male and high school/transfer), use the group chart to organize your information. First find the acceptable range of the total number of male applicants to be 193–218: 325 total minimum – 132 female = 193, and 350 total maximum – 132 female = 218. Since the question asks for the number of male high school applicants, plug the choices into your chart to solve for the number of male transfer students using the ratio. Then solve for the total number of male applicants to check if it is within the 193–218 range. In Choice (C), the number of male high school applicants is 156, divided by 3 since the male applicants are in a 3 to 1 ratio, and you get 52 male transfer applicants. 156 + 52 = 208 total male applicants, which is within the acceptable range. In Choice (D), the number of male high school applicants in now 162. Divide 162 by 3 to get 54 for the number of male transfer applicants. 162 + 54 = 216, which is also within the acceptable range. Choices (B), (E), and (F) are all either too big or too small. Choice (A) is a trap answer in that it is the number of female applicants, and Choice (G) is also incorrect since it represents the lower range of the total number of male applicants.

14. C

In this question, the order in which the friends are invited does not matter, so you're dealing with a combination.

The combination formula is: $C = \frac{n!}{r!(n-r)!}$, where n is the total number in the group you're drawing from and r is the number of those items you're selecting. With a total of 13 friends between them, Robin and Terry could invite $\frac{13!}{(5!)(8!)} = 1{,}287$ groups of 5 friends to their wedding. There are $\frac{7!}{(5!)(2!)} = 21$ groups that consist of only Robin's friends, and $\frac{6!}{(5!)(1!)} = 6$ groups that consist of only Terry's friends. Subtract these two numbers from 1,287, and find that there are 1,260 possible groups that contain at least one of Robin's friends and at least one of Terry's. Therefore, choice (C) is the correct answer.

15. A

To find the probability of exactly seven correct choices, think about all of the possible combinations of eight answers with one wrong (W) and seven right (R) choices. For example, Tony could get WRRRRRRR, RWRRRRRR, and so on. There are 8 different arrangements of one wrong and seven correct answers. The probability of choosing a correct answer for one question is $\frac{1}{2}$, and the probability of choosing a wrong answer is the same. Thus, the total number of possible outcomes is $\left(\frac{1}{2}\right)^8$. The probability of seven correct is the number of outcomes with seven correct divided by the total number of outcomes. The probability of exactly seven correct: $8 \times \frac{1}{2^8} = \frac{1}{32}$.

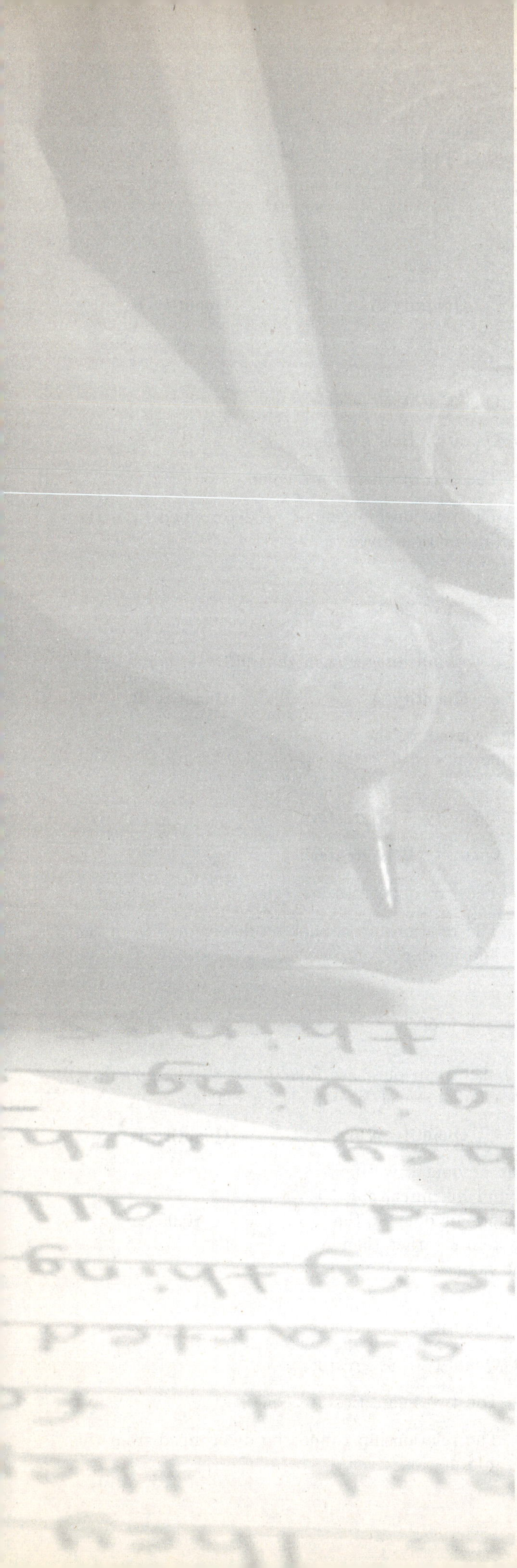

Chapter 9
Sample Section 1

Question 1 of 26

Quantity A	Quantity B
$\frac{9}{10} - \frac{8}{9}$	$\frac{8}{9} - \frac{9}{10}$

○ Quantity A is greater.

○ Quantity B is greater.

○ The two quantities are equal.

○ The relationship cannot be determined from the information given.

Question 2 of 26

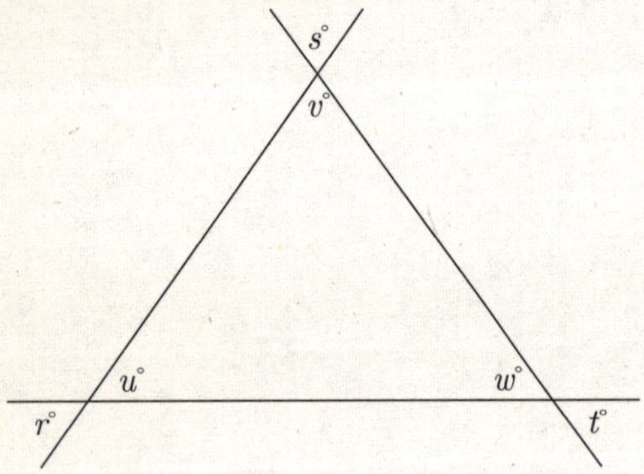

Quantity A	Quantity B
$r + t + v$	$s + u + w$

○ Quantity A is greater.

○ Quantity B is greater.

○ The two quantities are equal.

○ The relationship cannot be determined from the information given.

Question 3 of 26

$$x > 0$$
$$\frac{2x}{3} < 7$$

Quantity A	Quantity B
x^2	100

○ Quantity A is greater.

○ Quantity B is greater.

○ The two quantities are equal.

○ The relationship cannot be determined from the information given.

Question 4 of 26

k is an integer such that $9(3)^3 + 4 = k$

Quantity A	Quantity B
The average of the prime factors of k	16

○ Quantity A is greater.

○ Quantity B is greater.

○ The two quantities are equal.

○ The relationship cannot be determined from the information given.

Question 5 of 26

A bakery can purchase flour at \$1.89 per pound or at a bulk rate of 50 pounds for \$90.00.

Quantity A	Quantity B
The amount saved per pound by purchasing 50 pounds of flour at the bulk rate rather than by the pound	\$0.09

○ Quantity A is greater.

○ Quantity B is greater.

○ The two quantities are equal.

○ The relationship cannot be determined from the information given.

Question 6 of 26

$$x = ((24 \div 6) + 2) \times 5$$
$$y = (24 \div (6 + 2)) \times 5$$

Quantity A	Quantity B
$x - y$	$y - x$

- ○ Quantity A is greater.
- ○ Quantity B is greater.
- ○ The two quantities are equal.
- ○ The relationship cannot be determined from the information given.

Question 7 of 26

$$32 = |k|$$
$$31 = |k + 1|$$

Quantity A	Quantity B
k	32

- ○ Quantity A is greater.
- ○ Quantity B is greater.
- ○ The two quantities are equal.
- ○ The relationship cannot be determined from the information given.

Question 8 of 26

Alice weighs a kilograms, and Bob weighs b kilograms less than Alice.

$$ab \neq 0$$

Quantity A	Quantity B
The sum of Alice's weight and Bob's weight	$2a$ kilograms

- ○ Quantity A is greater.
- ○ Quantity B is greater.
- ○ The two quantities are equal.
- ○ The relationship cannot be determined from the information given.

Question 9 of 26

$$0 < xy$$

Quantity A	Quantity B
$\frac{5}{2x} - \frac{2}{2y}$	$\frac{5y - 2x}{2x - 2y}$

- ○ Quantity A is greater.
- ○ Quantity B is greater.
- ○ The two quantities are equal.
- ○ The relationship cannot be determined from the information given.

Question 10 of 26

If $x^* = x^2 + 4$, which of the following is equivalent to 2^* ?

- ○ $(-2)^*$
- ○ 0^*
- ○ 4^*
- ○ 6^*
- ○ 8^*

Question 11 of 26

Which of the following is greater than $\frac{2}{3}$ and less than $\frac{5}{6}$?

*Indicate **all** possible values.*

- ☐ $\frac{11}{18}$
- ☐ $\frac{13}{18}$
- ☐ $\frac{3}{4}$
- ☐ $\frac{7}{9}$
- ☐ $\frac{8}{9}$

Click on your choice(s).

Question 12 of 26

A manufacturer is designing two rectangular game boards. The width of the smaller board is one-half the width of the larger board, and the length of the smaller board is one-sixth the length of the larger board. If the smaller board has area N, then what is the difference between the areas of the game boards, in terms of N ?

- ○ $3N$
- ○ $9N$
- ○ $11N$
- ○ $12N$
- ○ $18N$

Question 13 of 26

If one manual weighs 400 grams, how many kilograms does a box of 48 manuals weigh? (1 kilogram = 1,000 grams)

- ○ 0.192
- ○ 1.92
- ○ 19.2
- ○ 192
- ○ 1,920

Question 14 of 26

What is the sum of the distinct positive even factors of 12 ?

[ ]

Click on the answer box, then type in a number. Backspace to erase.

Question 15 of 26

A computer store has fixed monthly operating costs of 5,000 dollars. The store buys each computer for 600 dollars and sells each computer for 900 dollars. What is the least number of computers the store must buy and sell each month to cover the fixed operating costs?

- ○ 14
- ○ 15
- ○ 16
- ○ 17
- ○ 18

Question 16 of 26

If x is equal to 49.5 percent of $\frac{11}{23}$ of 0.996, then

- ○ $0.20 < x < 0.25$
- ○ $0.25 < x < 0.30$
- ○ $0.30 < x < 0.35$
- ○ $0.35 < x < 0.40$
- ○ $0.40 < x < 0.45$

Question 17 of 26

Set $A = \{3, 1, 7, 5, 11, x\}$

If the median of Set A above is one less than the mode of Set A, which of the following is a possible value of x ?

- ○ 3
- ○ 5
- ○ 6
- ○ 7
- ○ 9

Question 18 of 26

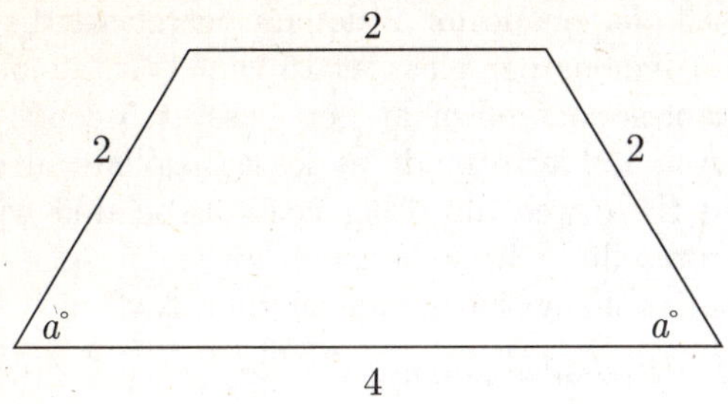

What is the area of the quadrilateral shown above?

- ○ $2\sqrt{3}$
- ○ $3\sqrt{3}$
- ○ 6
- ○ $6\sqrt{3}$
- ○ 8

Question 19 of 26

Anne's wage is \$10 per hour more than Mary's wage. Andy's wage is \$6 per hour more than Mark's wage. Mary's wage is \$2 per hour more than Mark's wage.

Quantity A	**Quantity B**
Anne's wage	Andy's wage

- ○ Quantity A is greater.
- ○ Quantity B is greater.
- ○ The two quantities are equal.
- ○ The relationship cannot be determined from the information given.

Question 20 of 26

M is the sum of three consecutive odd integers, the least of which is t. In terms of M, which of the following is the sum of three consecutive odd integers, the greatest of which is t ?

- ○ $M + 12$
- ○ $M + 6$
- ○ $M - 6$
- ○ $M - 12$
- ○ $\frac{M}{2}$

Question 21 of 26

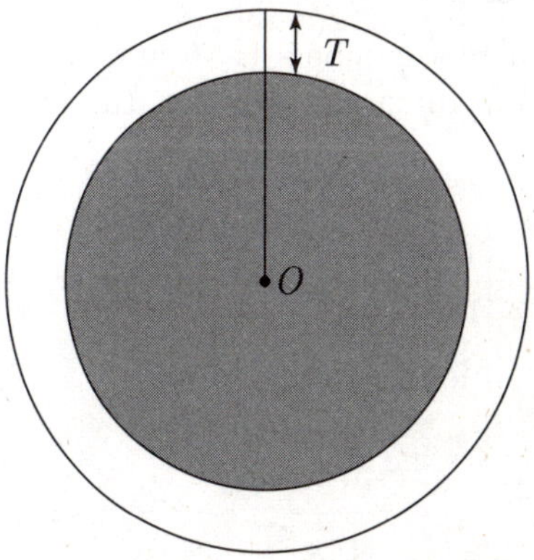

The figure above shows a circular garden plot with center O surrounded by a circular sidewalk, also with center O. If the area covered by the garden and sidewalk together is 169π and of the garden alone is 144π, what is the value of t ?

- ○ 1
- ○ 2
- ○ 3
- ○ 5
- ○ 12.5

Question 22 of 26

If the average (arithmetic mean) of a and b is 6, and $a^2 - b^2 = 2$, what is the value of $a - b$?

○ $\frac{1}{6}$

○ $\frac{1}{3}$

○ $\frac{1}{2}$

○ 2

○ 3

Question 23 of 26

If the sum of two numbers is 19 and their product is 88, what is the difference between the numbers?

○ 2

○ 3

○ 4

○ 5

○ 6

Question 24 of 26

If l is not equal to –3 or –4, then $\frac{l}{l+4} + \frac{3}{l+3} =$

○ 1

○ $\frac{7}{l+4}$

○ $\frac{l+3}{2l+7}$

○ $\frac{3l}{(l+4)(l+3)}$

○ $\frac{l^2+6l+12}{(l+4)(l+3)}$

Question 25 of 26

Pierre sells lemonade for 2 dollars per glass. His costs include 5 lemons per glass at 25 cents per lemon, and 3 tablespoons of sugar per glass at 5 cents per tablespoon. If Pierre made at least 5 dollars in one day, and there were no other costs associated with making or selling the lemonade, how many glasses of lemonade could he have sold on that day?

*Indicate **all** possible values.*

☐ 6

☐ 7

☐ 8

☐ 9

☐ 10

☐ 11

Click on your choice(s).

Question 26 of 26

Aquarium A holds 150 gallons more sea water than does Aquarium B, and both aquariums are currently filled to capacity. If 20 gallons are removed from each aquarium, then Aquarium A would hold four times as much sea water as Aquarium B holds. What is the combined total number of gallons of sea water that can be held in both aquariums?

○ 290

○ 250

○ 220

○ 200

○ 70

Chapter 10 Answers and Explanations to Sample Section 1

1. A

Use the Bow Tie to find the value of each quantity. Quantity A = $\frac{81-80}{90}=\frac{1}{90}$. Quantity B = $\frac{80-81}{90}=-\frac{1}{90}$. Quantity A is greater than Quantity B.

2. C

Vertical angles are equal, so $s = v$, $r = u$, and $t = w$. Therefore, the quantities are equal. To prove it, Plug In. Let $u = 50$, $v = 100$, and $w = 30$; the vertical-angle rule makes $s = 100$, $r = 50$, and $t = 30$, and now both $r + t + v$ and $s + u + w$ total 180.

3. D

First, find the value of x. Multiply both sides by 3 to get $2x < 21$. Divide both sides by 2 to find that $x < 10.5$. If $x = 2$, then Quantity B is greater because $2^2 = 4$. Eliminate choices (A) and (C). If $x = 10.2$, then Quantity A is greater because $(10.1)^2$ is greater than 10^2, or 100. Eliminate choice (B), and select choice (D).

4. C

Solve for k: $9(3)^3 + 4 = 9(27) + 4 = 243 + 4 = 247$. So, $k = 247$. Use trial and error to find the prime factors of 247. They are 13 and 19. The average of 13 and 19 is 16, so the quantities are equal.

5. C

First, divide $90 by 50 to find the bulk rate for a single pound of flour: $\frac{\$90}{50} = \$1.80$. The amount saved per pound is $.09. The quantities are equal.

6. A

Remember PEMDAS. $x = ((4) + 2) \times 5 = 6 \times 5 = 30$; $y = (24 \div (8)) \times 5 = 3 \times 5 = 15$. Quantity A = 15, and Quantity B = –15.

7. B

The solutions to the equation $32 = |k|$ are 32 and –32. Of those, only –32 works in the equation $31 = |k + 1|$. Quantity A has the value of –32, so Quantity B is greater.

8. B

Plug in. First, try $a = 100$ and $b = 2$: Alice weighs 100 kilograms, Bob weighs 98 kilograms, and the sum of their weights is 198 kilograms. Quantity B is 2 × 100, or 200. Quantity B is greater, so eliminate answer choices (A) and (C). Now, try a second set of numbers: If $a = 50$ and $b = 10$, then Quantity A is 90 and Quantity B is 100. Quantity B is, again, greater—as it will be with any set of numbers that meets the restriction ($ab \neq 0$).

9. D

Plug in for x and y. Because xy must be positive, either both variables are positive, or both are negative. First, let $x = 5$ and $y = 6$: Quantity A is $\frac{1}{3}$, and Quantity B is –10. Quantity A is greater, so eliminate answer choices (B) and (C). Now let $x = -\frac{1}{10}$ and $y = -\frac{1}{2}$. Quantity A is –23, and Quantity B is $-\frac{23}{8}$. Quantity B is now greater, so eliminate answer choice (A), and you're left with answer choice (D).

10. **A**

The weird little symbol (*) gives you a set of directions. $2^* = (2)^2 + 4 = 4 + 4 = 8$. Find the answer that also equals 8. $(-2)^* = (-2)^2 + 4 = 4 + 4 = 8$.

11. **B, C, and D**

Finding common denominators allows you to compare the values of fractions easily. Because two answer choices are expressed in eighteenths (and two in easily-convertible ninths) start there: $\frac{2}{3} = \frac{12}{18}$, and $\frac{5}{6} = \frac{15}{18}$. Eliminate answer choice (A), and select choice (B). Choice (D) converts to $\frac{14}{18}$, so it's correct; choice (E), though, converts to $\frac{16}{18}$ and can be eliminated. The answers are listed in ascending order, and so answer choice (C), falling between correct answer choices (B) and (D), must be correct as well. If time allows, verify by converting to twelfths: $\frac{2}{3} = \frac{8}{12}$, and $\frac{5}{6} = \frac{10}{12}$; choice (C), $\frac{3}{4}$, converts to $\frac{9}{12}$, and so, as expected, is correct.

12. **C**

Plug in. If the width of the smaller board is 3, then the width of larger board must be 6. If the length of the smaller board is 2, then the length of the larger board must be 12. The area of the smaller board is $3 \times 2 = 6$. Therefore, $N = 6$. The area of the larger board is $12 \times 6 = 72$. The question asks for the difference between the areas: $72 - 6 = 66$. The last step is to check which answer choice equals 66 when $N = 6$. Answer choice (C) works.

13. **C**

Because the answer choices are far apart, you can estimate. If one manual weighs 400 grams, then 48 manuals weigh 48×400, or about 20,000 grams. To convert grams to kilograms, divide by 1,000. $20{,}000 \div 1{,}000 = 20$. Look for an answer a little less than 20. Only choice (C) is close.

14. **24**

Start by listing the factors of 12 in pairs: 1 and 12, 2 and 6, and 3 and 4. Because the question asks for the sum of the distinct positive even factors, cross off 1 and 3. Now, add up what's left: $2 + 4 + 6 + 12 = 24$.

15. **D**

The store makes a profit of $300 for each computer sold. Plug in the answer choices, starting with answer choice (C). 16 computers multiplied by $300 yields a profit of $4,800, not enough to cover costs. You need to sell more computers, so eliminate answers choices (A), (B), and (C). Try choice (D). 17 computers multiplied by $300 yields a profit of $5,100, just enough to cover costs.

16. **A**

Estimate. If x is less than 50 percent of a fraction that is less than $\frac{1}{2}$ of a number that is less than 1, then x must be less than $\frac{1}{4}$ $(\frac{1}{2} \times \frac{1}{2} \times 1 = \frac{1}{4})$. Therefore, choice (A) is correct.

17. D

Each number in the set occurs only once, and the mode is the most frequently occurring number. To create a mode, x must be same as one of the other numbers in the set. Eliminate choice (C) and choice (E). Plug in the answers. For choice (B), the mode is 5. The median of {1, 3, 5, 5, 7, 11} is 5. The median is equal to the mode. You need a larger number for x. For choice (D), the mode is 7 and the median of {1, 3, 5, 7, 7, 11} is $\frac{(5+7)}{2} = 6$. The mode is now one less than the median.

18. B

Redraw this figure by adding two descending lines from the upper corners of the figure perpendicular to the base below so that you have a rectangle and two right triangles. The triangles each have a base of 1 and a hypotenuse of 2. Use the Pythagorean theorem or the ratio for 30-60-90 triangles ($a : a\sqrt{3} : 2a$) to find the height is $\sqrt{3}$. The area of each triangle is $\frac{1}{2}(1)\left(\sqrt{3}\right)$. The area of the rectangle is $(2)\left(\sqrt{3}\right)$. The sum of the rectangle and two triangles is $2\sqrt{3} + 2\left(\frac{1}{2}\sqrt{3}\right) = 2\sqrt{3} + \sqrt{3} = 3\sqrt{3}$.

19. A

If Mary's wage is \$2 more than Mark's, and Andy's wage is \$6 more than Mark's, then Andy's wage is \$4 more than Mary's. Anne's wage is \$10 more than Mary's, so Anne's wage is \$6 more than Andy's. Quantity A is greater than Quantity B.

20. D

Plug in, and let $t = 5$. $M = 5 + 7 + 9 = 21$—the sum of the three consecutive odd integers, of which t is the smallest. For the second set, t is the greatest integer in the series. The sum of the new series is $5 + 3 + 1 = 9$. Plug $M = 21$ into the answers, and answer choice (D) matches the target of 9.

21. A

Two circles share a center. The bigger circle that includes the garden and the sidewalk has an area of 169π. The radius of the large circle with the garden and sidewalk is 13. The garden plot has area 144π, so the radius of the small circle is 12. Because the circles share an origin, the width of the sidewalk, t, is the difference between the radii, or $13 - 12 = 1$. Therefore, answer choice (A) is correct.

22. A

Factor the quadratic expression. $a^2 - b^2$ becomes $(a + b)(a - b)$. The average of two numbers is 6, so their sum, $(a + b)$, is $2 \times 6 = 12$. Substitute the value to get: $12(a - b) = 2$. Thus, $a - b = \frac{1}{6}$.

23. B

Plug in numbers that have a sum of 19, and check to see if their product is 88. For example, the sum of 9 and 10 is 19, but their product is 90. 9 and 10 can't be the two numbers. Try another pair, such as 8 and 11. Their sum is 19, and their product is 88. These numbers match the information given in the problem, and their difference is 3.

24. E

Plug in. If $l = 2$, then $\frac{l}{l+4} + \frac{3}{l+3} = \frac{2}{6} + \frac{3}{5} = \frac{10+18}{30} = \frac{28}{30} = \frac{14}{15}$. The target is $\frac{14}{15}$. Plug 2 for l into the answer choices. Answer choice (E) is the only answer choice that matches the target.

25. D, E, and F

First, figure out Pierre's per-glass profit. His per-glass revenue is $2; his per-glass costs are 25 cents × 5 lemons per glass = 125 cents per glass for lemons, and 5 cents × 3 tablespoons of sugar per glass = 15 cents per glass for sugar, for a total of 140 cents per glass. Pierre's profit on each glass is 60 cents. At a profit of 60 cents per glass, Pierre must sell at least 9 glasses of lemonade to make 5 dollars. Each answer choice greater than or equal to 9 is correct.

26. A

Plug in the answers, starting with choice (C). If the combined total was 220 and *A* is 150 more than *B*, then *A* is 75 more than half of 220, and *B* is 75 less than half of 220. So, *A* is 185 and *B* is 35. Remove 20 from each to get *A* is 165 and *B* is 15. Is 165 four times 15? No, so choice (C) is not correct. It's hard to tell whether choice (C) was too large or too small, so just pick a direction. For choice (A), if the combined total was 290, and *A* is 150 more than *B*, then *A* is 75 more than half of 290, and *B* is 75 less than half of 290. So, *A* is 220 and *B* is 70. Remove 20 from each to get *A* is 200 and *B* is 50. Is 200 four times 50? Yes, so choice (A).

Chapter 11
Sample Section 2

Question 1 of 26

The average (arithmetic mean) of a, $2a$, and 9 is 7.

Quantity A	Quantity B
5	a

○ Quantity A is greater.

○ Quantity B is greater.

○ The two quantities are equal.

○ The relationship cannot be determined from the information given.

Question 2 of 26

$$12 > 3x - 9y$$

Quantity A	Quantity B
6	$x - 3y$

○ Quantity A is greater.

○ Quantity B is greater.

○ The two quantities are equal.

○ The relationship cannot be determined from the information given.

Question 3 of 26

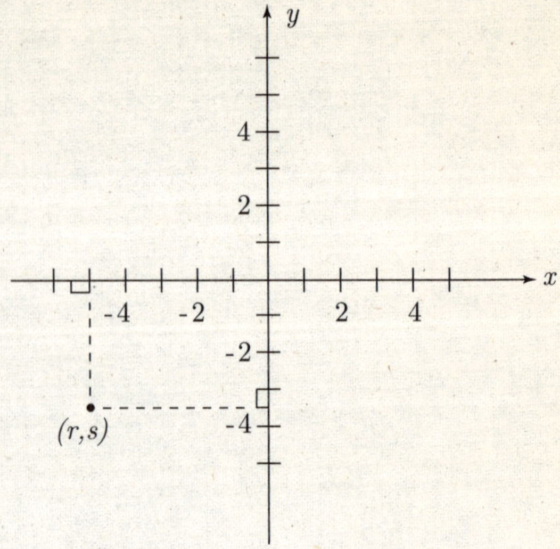

Quantity A	Quantity B
r	s

○ Quantity A is greater.

○ Quantity B is greater.

○ The two quantities are equal.

○ The relationship cannot be determined from the information given.

Question 4 of 26

$$\begin{array}{r} 145.3 \\ \underline{-AB.A} \\ 66.6 \end{array}$$

In the correctly calculated subtraction problem shown above, a and b represent digits from 0 to 9, inclusive.

Quantity A	Quantity B
15	$A + B$

○ Quantity A is greater.

○ Quantity B is greater.

○ The two quantities are equal.

○ The relationship cannot be determined from the information given.

Question 5 of 26

For all integers a and b, $a \# b = -|a + b|$

Quantity A	Quantity B
$(-10) \# 7$	$7 - 10$

- ○ Quantity A is greater.
- ○ Quantity B is greater.
- ○ The two quantities are equal.
- ○ The relationship cannot be determined from the information given.

Question 6 of 26

Company X sold 300 products this year. It sold twice as many of Product A as it did of Product C, and half of the products sold were Product B. Company X sold no products other than A, B, and C.

Quantity A	Quantity B
The number of Product A sold	100

- ○ Quantity A is greater.
- ○ Quantity B is greater.
- ○ The two quantities are equal.
- ○ The relationship cannot be determined from the information given.

Question 7 of 26

$z \neq 0$

Quantity A	Quantity B
$\frac{99z}{100}$	$\frac{100}{99z}$

- ○ Quantity A is greater.
- ○ Quantity B is greater.
- ○ The two quantities are equal.
- ○ The relationship cannot be determined from the information given.

Question 8 of 26

In a garden, there are only red, yellow, and blue flowers. One-third of the flowers are red, and 40 percent are blue. One flower is chosen at random.

Quantity A	Quantity B
The probability that the flower chosen is not red	The probability the flower chosen is not yellow

- ○ Quantity A is greater.
- ○ Quantity B is greater.
- ○ The two quantities are equal.
- ○ The relationship cannot be determined from the information given.

Question 9 of 26

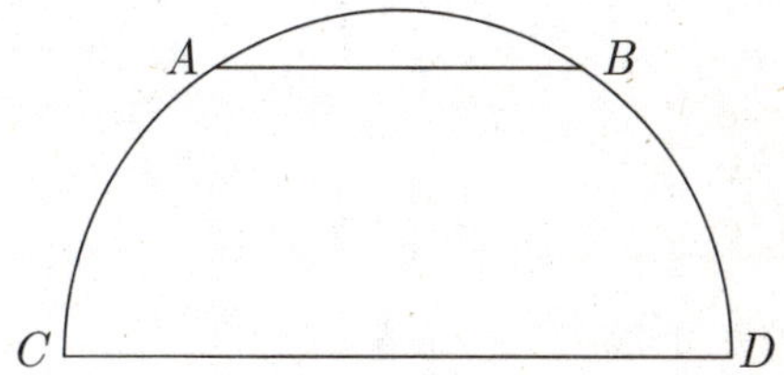

In the semicircle above, the length of arc AC is equal to the length of arc BD, and the length of arc AB is less than the length of arc BD.

Quantity A	Quantity B
$\frac{\text{the length of chord } AB}{\text{the length of chord } CD}$	$\frac{1}{2}$

- ○ Quantity A is greater.
- ○ Quantity B is greater.
- ○ The two quantities are equal.
- ○ The relationship cannot be determined from the information given.

Question 10 of 26

What is the value of $(4 + a)(4 - b)$ when $a = 4$ and $b = -4$?

- ○ –64
- ○ –16
- ○ 0
- ○ 16
- ○ 64

Question 11 of 26

If the average (arithmetic mean) of two numbers is 35 and one of the numbers is k, then what is the other number in terms of k ?

- ○ $35 + k$
- ○ $35 - k$
- ○ $35 - 2k$
- ○ $70 - k$
- ○ $70 - 2k$

Question 12 of 26

Alejandro calculated 25 percent of x instead of a 25 percent increase in x. Which of the following operations could Alejandro perform on his solution to produce the correct solution?

*Indicate **all** possible answers.*

- ☐ Add x
- ☐ Multiply by x
- ☐ Divide by 0.25
- ☐ Multiply by 5

Click on your choice(s).

Question 13 of 26

If the ratio of $x : y$ is $6 : 7$, and x is equal to 15, then the value of y is

- ○ 7
- ○ 13.5
- ○ 17
- ○ 17.5
- ○ 20

Question 14 of 26

If $x \neq 0$, what is the value of b when $x = a(b + c)$?

- ○ $c - \frac{a}{x}$
- ○ $\frac{x}{a} - c$
- ○ $ax - c$
- ○ $ax + c$
- ○ $x - ac$

Question 15 of 26

$$\frac{3^6 + 3^4 + 3^2}{3^2 + 3^4 + 3^6}$$

[ ]

Click on the answer box, then type in a number. Backspace to erase.

Question 16 of 26

If $k = 6 \times 17$, then which of the following is a multiple of k ?

- ○ 68
- ○ 78
- ○ 85
- ○ 136
- ○ 204

Question 17 of 26

The ratio of the degree measures of the angles of a triangle is 2 : 3 : 4. Which of the following is the sum of the degree measures of the smallest and largest angles?

- ○ 40°
- ○ 80°
- ○ 100°
- ○ 120°
- ○ 140°

Question 18 of 26

The price of a slice of pizza is $(8a + b)$ cents, the price of a soda is $(8b + a)$ cents, and the sum of the two prices is \$1.35.

Quantity A	**Quantity B**
a	b

- ○ Quantity A is greater.
- ○ Quantity B is greater.
- ○ The two quantities are equal.
- ○ The relationship cannot be determined from the information given.

Question 19 of 26

If $y = \sqrt{0.36x^8}$, then $y =$

- ○ $0.06x^2$
- ○ $0.06x^4$
- ○ $0.06x^5$
- ○ $0.6x^2$
- ○ $0.6x^4$

Question 20 of 26

If $x^2 + 2xy + y^2 = 25$, then $(x + y)^3$ could be

- ○ 5
- ○ 15
- ○ 50
- ○ 75
- ○ 125

Question 21 of 26

A grain elevator moves 200 pounds of grain every 15 minutes. If the elevator starts moving grain at noon, and stops between 3:00 P.M. to 4:00 P.M., which of the following could be the weight, in pounds, of the grain moved by the elevator?

*Indicate **all** possible answers.*

- ☐ 800
- ☐ 2,400
- ☐ 2,732
- ☐ 3,000
- ☐ 3,200
- ☐ 3,600

Question 22 of 26

If $a = (0.404)^3$, $b = \sqrt[3]{0.404}$, and $c = 0.404$, then which of the following is true?

- ○ $a > b > c$
- ○ $b > a > c$
- ○ $a > c > b$
- ○ $c > a > b$
- ○ $b > c > a$

Question 23 of 26

At least 15 percent of the students registered for Professor Tyler's course dropped out before the end of the course. The drop-out rate for Professor Quin's course was 20 percent greater than that for Professor Tyler's course. If 120 students remained on the last day of Professor Quin's class, which of the following could be the number of students who signed up for Professor Quin's class?

*Indicate **all** possible values.*

- ☐ 42
- ☐ 135
- ☐ 142
- ☐ 146
- ☐ 147
- ☐ 162
- ☐ 184
- ☐ 185

Click on your choice(s).

Question 24 of 26

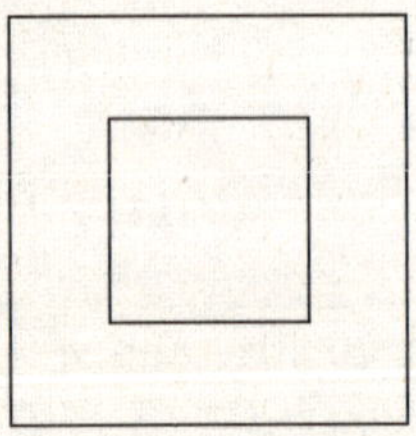

In the figure above, if the area of the larger square region is twice the area of the smaller square region, and if a diagonal of the smaller square has a length of 1 foot, then a side of the larger square is how many feet longer than a side of the smaller square?

- ○ $\frac{2-\sqrt{2}}{2}$
- ○ $\sqrt{2}-1$
- ○ 1
- ○ $\sqrt{2}$
- ○ 2

Question 25 of 26

If $5a - 3b = -9$ and $3a + b = 31$, then $a + b =$

- ○ –7
- ○ –3
- ○ 6
- ○ 13
- ○ 19

Question 26 of 26

Claire estimates her chances of being admitted to her top five schools as follows: 80 percent for Debs U, 70 percent for Powderly State, 50 percent for Randolph A&M, 20 percent for Reuther, and 10 percent for Chavez Poly. If Claire's estimates are correct, her chance of being admitted to at least one of her top five schools is between

- ○ 75% and 79.99%
- ○ 80% and 84.99%
- ○ 85% and 89.99%
- ○ 90% and 94.99%
- ○ 95% and 99.99%

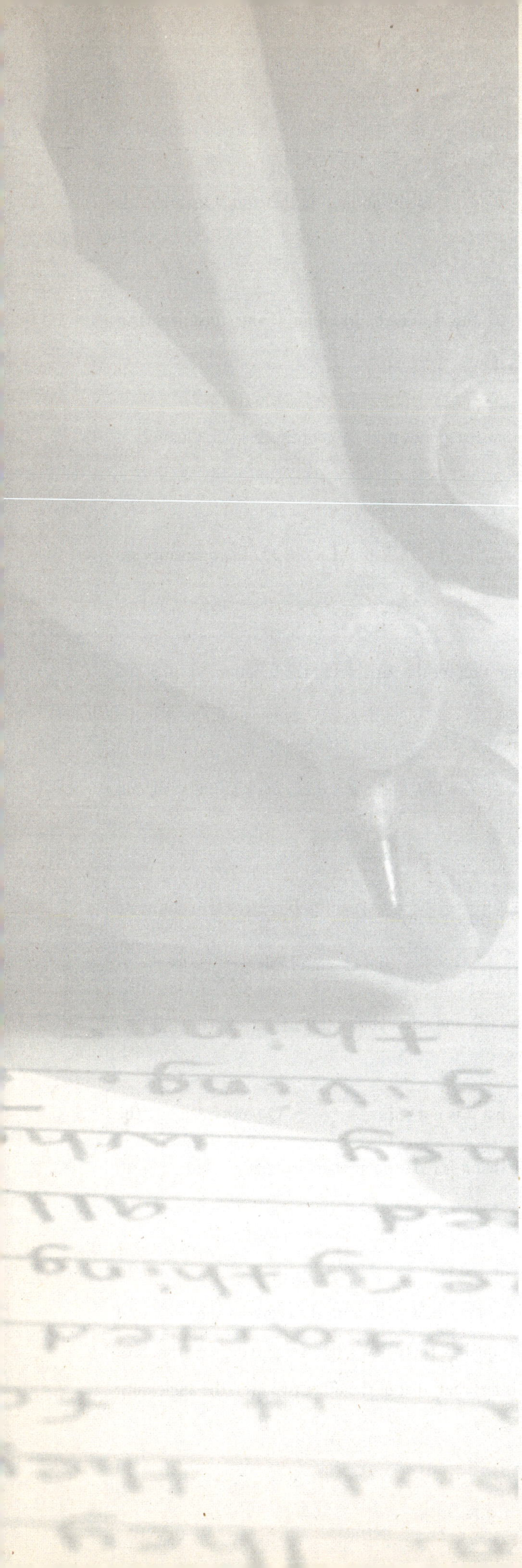

Chapter 12
Answers and Explanations to Sample Section 2

1. A

If the average of 3 things is 7, the total is $(3 \times 7) = 21$. $21 = a + 2a + 9 = 3a + 9$, and $a = 4$. Therefore, Quantity A is greater.

2. A

Divide both sides of the inequality by 3 to get $4 > x - 3y$. Quantity B must be less than 4, and, thus, less than Quantity A.

3. B

Follow the dotted lines. r corresponds to the x-coordinate, –5. s corresponds to the y-coordinate, –3.5. Quantity B is greater than Quantity A.

4. C

Try rearranging the problem. If $\frac{\begin{array}{r}145.3\\-AB.A\end{array}}{66.6}$, then $\frac{\begin{array}{r}145.3\\-66.6\end{array}}{AB.A}$. Hence, $AB.A = 78.7$, and $A + B = 15$. The quantities are equal.

5. C

For Quantity A, substitute the numbers the provided in the question for the variables in the function: $-|-10 + 7| = -|-3| = -3$. Quantity B is –3. The quantities are equal.

6. C

The company sold $300 \div 2 = 150$ of Product B. The remainder, $A + C = 150$. Because $A = 2C$, $3C = 150$. $C = 50$, and $A = 100$. The quantities are equal.

7. D

Plug in 1 for z. Quantity A is $\frac{99 \times 1}{100} = \frac{99}{100}$, and Quantity B is $\frac{100}{99 \times 1} = \frac{100}{99}$. Quantity B is greater, so eliminate answer choices (A) and (C). Plug in –1 for z. Quantity A is $\frac{99 \times -1}{100} = -\frac{99}{100}$, and Quantity B is $\frac{100}{99 \times -1} = -\frac{100}{99}$.

Quantity A is closer to 0, and, thus, is greater. Eliminate choice (B), and select choice (D).

8. B

The probability that the flower is red is $\frac{1}{3}$, so the probability that it is not red is $1 - \frac{1}{3} = \frac{2}{3}$. Quantity A is $\frac{2}{3}$. The probability that the flower is not yellow equals the probability that it is red + the probability that it is blue. $\frac{1}{3}$ + 40 percent $= \frac{1}{3} + \frac{2}{5} = \frac{11}{15}$. Quantity B is $\frac{11}{15}$ and is greater.

9. B

Redraw the figure so that arc AB looks much smaller than BD. Now, draw radii from the center of the circle to A and to B. If the new angle were 90 degrees, Quantity A would equal $\frac{1}{2}$. If the new angle were 60 degrees, the three arcs would have equal lengths. Because the length of arc AB is less than that of the other two arcs, the angle must be less than 60 degrees. Quantity A must be less than $\frac{1}{2}$.

10. E

You are given a value for each of the variables. If you put these values directly into $(4 + a)(4 - b)$, you get $(4 + 4)(4 - (-4)) = (8)(4 + 4) = (8)(8) = 64$.

11. D

If the average of 2 numbers is 35, the total is $(2 \times 35) = 70$. Plug in for k, and let $k = 20$. $70 - 20 = 50$, the other number and the target answer. Go to the answer choices and plug in 20 for k. Only answer choice (D) matches your target of 50.

12. A and D

If x is 100, then Alejandro found 25 percent of 100 = 25 instead of increasing 100 by 25 to 125. The question asks what Alejandro could do to turn 25 into 125. In math terms, Alejandro found $0.25x$ instead of $1.25x$. If he adds x, this will work because $25 + 100 = 125$, or in math terms $0.25x + 1x = 1.25x$. Multiplying by x will not work because $25 \times 100 = 2{,}500$, not 125. Dividing by 0.25 will not work because $25 \div 0.25 = 100$, not 125. Multiplying by 5 works because $25 \times 5 = 125$, and $0.25x \times 5 = 1.25x$.

13. D

Use the ratio box. Fill in that the ratio of x to y is 6 to 7. Fill in that the actual value of x is 15. The multiplier is $\frac{15}{6} = \frac{5}{2}$. Use this multiplier to find y: $7 \times \frac{5}{2} = \frac{35}{2} = 17.5$.

14. B

Plug in. If $b = 2$, $c = 3$, and $a = 4$, then $x = 20$. The question asks for the value of b, so the target answer is 2. Plug $c = 3$, $a = 4$, and $x = 20$ into all the answers to find the answer that yields 2. Choice (A) yields a fraction, not 2. Choice (B) yields $\frac{20}{4} - 3 = 2$, so keep this answer. Estimate that choices (C) and (D) are larger than 2. Choice (E) yields 8, not 2. Choice (B) is the only choice that yielded the target answer.

15. 1

Notice that the exact same numbers appear in the numerator and denominator of the fraction. Because it doesn't matter in what order you add quantities, the numerator and denominator of the fraction are the same. You don't even have to add it up. The answer is 1 because a nonzero number divided by itself is 1.

16. E

You could multiply 6 by 17 and start dividing the answer choices by the result, but there's an easier way. Because $6 = 3 \times 2$, the correct answer must be a multiple of 2. Eliminate choice (C) because it's odd. Next, the correct answer is a multiple of 3. To check to see which numbers are multiples of 3, add their digits, and check to see if the sum is a multiple of 3. For example, choice (A) is $6 + 8 = 14$, which is not a multiple of 3. Eliminate choice (A). You can also eliminate choice (D) this way. Try dividing choice (B) by 17. Because you don't get an integer, that leaves only choice (E).

17. D

There are 180 degrees in a triangle, so you know that $2x + 3x + 4x = 180$. Solve for x: $9x = 180$, and $x = 20$. The smallest angle is 40 degrees, the middle angle is 60 degrees, and the largest angle is 80 degrees. The sum of the smallest and largest angles is 120, or choice (D).

18. D

You only know the sum of the two prices: $(8a + b) + (8b + a) = 135$ cents, or $9a + 9b = 135$. You could simplify this by dividing by 9 to get $a + b = 15$. However, you cannot determine the values of a and b individually, nor their values relative to each other. For example, try $a = 5$ and $b = 10$, and then try $a = 10$ and $b = 5$. Different outcomes are possible, and the answer is choice (D).

19. E

Work with numerical portion of the square root first $\sqrt{0.36} = \frac{\sqrt{36}}{\sqrt{100}} = \frac{6}{10} = 0.6$. Eliminate answer choices (A), (B), and (C). Next, work with the variable. If you are struggling with exponent rules you can expand it out. $\sqrt{x^8} = \sqrt{x \times x \times x \times x \times x \times x \times x \times x} = x^4$. So $y = 0.6\,x^4$.

20. E

Notice that $x^2 + 2xy + y^2 = 25$ is a quadratic equation. The question asks for $(x + y)^3$. The first step is to factor $x^2 + 2xy + y^2 = 25$. It becomes $(x + y)\,(x + y) = 25$. Therefore, $(x + y) = \pm 5$. Because you know that $(x + y)$ could be 5 or -5, you know $(x + y)^3$ could be 125 or -125. The correct answer is (E).

21. B, C, D, and E

Because the elevator moves 200 pounds of grain every 15 minutes, it moves 800 pounds every hour. Starting at noon and stopping at 3:00, the elevator moves $3 \times 800 = 2{,}400$ pounds of grain. Eliminate answer choice (A). At the opposite extreme, starting at noon and stopping at 4:00, the elevator moves $4 \times 800 = 3{,}200$ pounds. Eliminate answer choice (F). The upper and lower bounds having been established, any answer choices that fall on or between them—that is, choices (B), (C), (D), and (E)—are correct.

22. E

Remember that a fraction or decimal between 0 and 1 gets smaller when you cube it. So, $c > a$. Conversely, it gets bigger when you take the cube root, so $b > c$. Therefore, $b > c > a$.

23. E, F, G, and H

If students are dropping out of the course, then there had to have been more than 120 students originally. Eliminate answer choice (A). Next, find Professor Quin's drop-out rate: 20 percent (or $\frac{1}{5}$) of 15 is 3, so Professor Quin's rate is 3 greater than 15 percent, or 18 percent. If 18 percent of the students dropped out over the length of the course, then 82 percent remain on the last day. 82 percent of x is 120: $\frac{82}{100} \times x = 120$. The value of x is 146.34; because you can't have .34 of a person, the total must have been 147. The rate is at least 18 percent, so there could have been more students originally; hence, any answer choices that are greater than 147 can be correct as well.

24. A

First, find a side of the smaller square. $x^2 + x^2 = 1$. $2x^2 = 1$. $x^2 = \frac{1}{2}$. $x = \frac{\sqrt{2}}{2}$. Alternatively, you can apply the ratio for the sides of an isosceles right triangle ($s : s : s\sqrt{2}$) to solve for the length of a side. The area of this square is $\left(\frac{\sqrt{2}}{2}\right)^2$ or $\frac{1}{2}$. Therefore, the area of the larger square must be 1. The formula for the area of a square is $a = s^2$. $s^2 = 1$, which means $s = 1$. The difference between the larger and smaller sides is $1 - \frac{\sqrt{2}}{2} = \frac{2-\sqrt{2}}{2}$.

25. E

Multiply both sides of the second equation by 3 to get $9a + 3b = 93$. Add the first and second equations:

If $5a - 3b = -9$

$+ \underline{9a + 3b = 93}$

$14a = 84$

Solve to find that $a = 6$. Plug $a = 6$ back into $5a - 3b = -9$ to find the value of b: $5(6) - 3b = -9$, or $30 - 3b = -9$, and $b = 13$. The question asks for $a + b$, which is $6 + 13 = 19$.

26. E

To figure out the probability of being accepted to at least one university, use the following formula: P(at least one) = 1 – P(none). The chance that Claire gets into none of her top five schools is $0.2 \times 0.3 \times 0.5 \times 0.8 \times 0.9$, which is about 2 percent. Therefore, her chance of getting into at least one of her top five schools is about 98 percent.

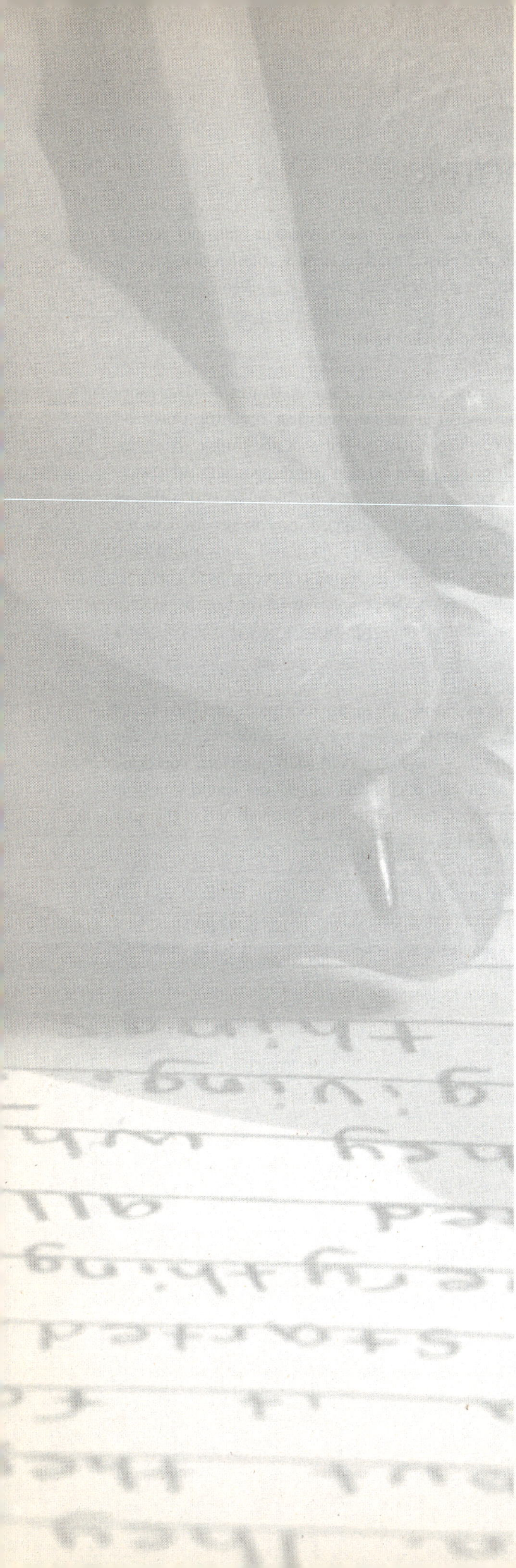

Chapter 13
Triggers and Glossary

Trigger, Response

You may have noticed that the hardest part of any math question is simply getting started. Once you know what to do, the actual math is simply about writing down numbers and occasionally using the calculator. But when a question appears on the screen, most people's first impulse is to keep rereading the question, and then stare off into the distance and think about what to do.

Remember that the math section of the GRE is not about thinking. The more time you spend staring at a problem without actually writing anything down on your scratch paper, the more time you are wasting. Triggers are simple things to look for in GRE math problems: Once you see a certain trigger, you should always have the same response. That way, rather than trying a problem several different ways before realizing that you could have simply plugged in, you get used to seeing the problem, noticing variables in the answers (Trigger) and knowing to Plug In (Response). Math problems on the GRE test the same concepts, over and over again, generally in the same couple of ways. When you're ready for the GRE's repetitive questions, the slight variations between questions won't throw you for a loop.

The best way to practice Triggers is to simply open up to a page of GRE math questions, perhaps from this book, from *Cracking the New GRE,* or from *The Official Guide to the revised GRE,* and look for Triggers in each question. You don't even have to solve any questions (although it couldn't hurt), just spend a couple minutes per page looking through the questions, telling yourself what the first couple of steps for each question would be.

These Triggers are listed in the order in which they appear in the book. If you are unfamiliar with the question types that use a particular Trigger, go back to that chapter to see questions of that type, how we solved them, and what clues we found that told us how to solve them.

Chapter 3: The Nuts and Bolts

Trigger: Problem contains percentages.
Response: Translate, Convert percentages to fractions, use Proportions, or use Tip Calculation.

Trigger: Question asks for "percent change," "percent increase," or "percent decrease."
Response: Write the percentage change formula: Percent change = $\frac{change}{original} \times 100$

Trigger: Exponent problems with large numbers.
Response: Factor the base to compare exponents. Use MADSPM.

Chapter 4: Algebra, and How to Get Rid of It

Trigger: Variables in the answer choices.
Response: Plug in.

Trigger: Variables in the answer; Problem says "must be"
Response: Plug in a simple number, then use ZONEF numbers.

Trigger: Quant Comp with variables.
Response: Set up your scratch paper and plug in using ZONEF

Trigger: "How much," "How many," "What is the value," numbers in the answer choices.
Response: Plug in the Answers (PITA)

Chapter 5: Math in the Real World

Trigger: The word "average."
Response: Draw an average pie for every time the word average appears in the question.

Trigger: The word "median" appears in the problem.
Response: Put the list of numbers in order and find the middle number.

Trigger: The word "ratio" appears in the problem.
Response: Draw a Ratio Box on your scratch paper.

Chapter 6: Geometry

Trigger: Two parallel lines cut by a transversal.
Response: Label all acute (small) angles as equal, and all obtuse (large) angles as equal.

Trigger: Need to know the side of a right triangle.
Response: Write down $a^2 + b^2 = c^2$ and plug in the two sides you know.

Trigger: Triangle question contains the word "area."

Response: Write down $A = \frac{1}{2}bh$ and plug in what you know. The height will always be perpendicular to the base.

Trigger: Problem with parallelogram, rectangle, or square contains the word "area."
Response: Write down the area formula and plug in information.

Trigger: Problem mentions "perimeter."
Response: Find the length of each side and add up all sides.

Trigger: Circle problem contains the word "circumference."
Response: Write $C = 2\pi r$ or $C = \pi d$ on your scratch paper.

Trigger: Circle problem contains the word "area."
Response: Write $A = \pi r^2$ on your scratch paper.

Chapter 7: The Rest of the Story

Trigger: Factorials with division.
Response: Expand factorial and reduce.

Trigger: Factorials with addition or subtraction.
Response: Factor out common factorials.

Trigger: The word "probability" appears in the problem.
Response: For each event, find the number of outcomes you want, and divide by the total number of outcomes. $\frac{want}{total}$

Trigger: Probability question asks "at least."
Response: Find the probability event won't happen, and subtract from 1.

Trigger: Group problem with overlap.
Response: Write down group formula: Total = [Group 1] + [Group 2] – [Both] + [Neither]

Trigger: Group question with no "Both Group A and B" elements.
Response: Draw Group Table:

	Group X	Group Y	Total
Group A			
Group B			
Total			

Trigger: Question contains the words "normal distribution" or "standard deviation."
Response: Draw a bell curve and label the mean and the 34-14-2 points for each standard deviation.

Trigger: The phrases "arrangements," "permutations," "combinations," "different ways," "many ways," or "different groups" appear in the problem.
Response: Draw a horizontal line for each choice we have to make.

Trigger: The words "team," "groups," "combinations," or order doesn't matter.
Response: At the end of the problem, divide answer by the factorial of the number of slots.

Glossary

Here is a list of mathematical terms that every GRE student should know well. Terms in *italics* have been cross-referenced from other definitions within this list.

A

Absolute Value
The absolute value of a number is defined as the distance of that number from zero. For the GRE, the important parts of absolute value are that the absolute value of a positive number is positive, the absolute value of a negative number is positive, and if a variable has an absolute value sign around it then that variable has two solutions: If $|x| = 5$ then $x = 5$ or -5.

Acute Angle
An angle that measures less than 90°.

All That Apply
A GRE question format that has square answer boxes. You must select every answer choice (out of anywhere between 3 and 8 answer choices) that applies. There is no partial credit; if any correct answer choices are not selected, or any incorrect answer choices are selected, the entire response is considered incorrect.

Altitude
The distance between the vertex and the *base* of a triangle, also known as the height of the triangle.

Arc
Any measurement around the circumference of a circle.

Area
The amount of space within a two-dimensional figure. The important formulas for area are: Triangle Area = $\frac{1}{2}bh$; Parallelogram or Rectangle Area = bh; Square Area = s^2; and Circle Area = πr^2

B

Ballparking
Approximating what the right answer might be and eliminating all impossible answer choices before doing any calculations.

Base (of an exponent)
The bottom, larger number in an exponential expression. In the expression 3^4, the base is 3.

Base (of a triangle)
The bottom side of a triangle; used to find a triangle's area.

Binomial
An algebraic expression that contains two terms, such as $(x + 2)$.

Bisect
To cut into two equal parts.

C

Chord
A line segment that connects two points on a circle's *circumference*. The longest possible chord, the *diameter*, goes through the center of the circle.

Circumference
The *perimeter* of a circle. $C = 2\pi r$

Coefficient
A number that appears next to a *variable* and should be multiplied by that variable. In the expression $5a$, which is shorthand for "$5 \times a$," the coefficient is 5.

Combination
A possible arrangement of a certain number of terms when the order in which those items are selected does not matter. (See also *Permutation.*)

Concentric
Having the same center. (Most often used in terms of circles.)

Constant
Any number that is not a *variable.*

D

Denominator
The bottom number in a fraction. If a denominator equals zero, the fraction is undefined.

Diameter
A line segment that connects two points on a circle's *circumference* and goes through the center. The diameter is the circle's largest *chord,* and it is twice as long as the *radius.*

Difference
The result of subtraction. The difference of 5 and 3 is 2.

Distributive Property
A mathematical property whereby any number multiplied by a *sum* or *difference* of two or more numbers must be multiplied by all the numbers therein. The expression $3(2x + 5)$ can be rewritten, or distributed, to $(3 \times 2x) + (3 \times 5)$, or $6x + 15$.

Divisible
A number a is divisible by another number b if b divides into a evenly, with no remainder. In other words, $\frac{a}{b}$ is an *integer.*

Divisor
A number that can be divided into another number.

E

Equilateral Triangle
A triangle with three sides of the same length and three angles of the same measure (60°).

Exponent
The small number in the upper-right corner of an exponential expression. In the expression 3^4, the exponent is 4. This means the base 3 must be multiplied by itself 4 times $(3 \times 3 \times 3 \times 3)$.

Extra Information
Information in a math problem that doesn't actually help solve the problem. In the GRE, questions rarely have extra information. If you are ever stuck on a GRE problem, reread the question to see if there's any information you haven't used yet.

Even
Divisible by 2.

F

Factor
Any number that can be divided evenly into another number. Because 3 divides evenly into 12, 3 is a factor of 12.

Factorial
A process whereby an *integer* is multiplied by each of the positive integers less than itself exactly once. "Eight factorial" is denoted as 8! and can be found by multiplying $8 \times 7 \times 6 \times 5 \times 4 \times 3 \times 2 \times 1$, which equals 40,320.

F.O.I.L.
Acronym for First, Outside, Inside, Last that indicates how two *binomials* can be multiplied together.

Function
A method of showing a relationship between two or more variables. On the GRE, function questions may involve symbols that are not typically in math questions, and will require following the directions given for that particular function.

I

Improper Fraction
A fraction in which the *numerator* is greater than the *denominator.* These fractions can be converted to mixed fractions.

Integer
Any number that has no fraction or decimal associated with it. Think of integers as the counting numbers on the number line including 0. Integers can be positive or negative.

Irrational Number
Any number that cannot be denoted as a fraction. Most irrational numbers you'll come in contact with on the GRE will be square roots and π.

Isosceles Triangle
A triangle with two sides of the same length and two angles of the same measure.

L

Like Terms
Any algebraic terms that contain the exact same configuration of variables and therefore can be combined. For example, $3a^2b$ and $6a^2b$ are like terms, so they can be added to make $9a^2b$. The variables in $5xy^3$ and $10x^3y$ are similar but *not* identical, so these are not like terms.

M

MADSPM
A mnemonic device to remember the rules for combining exponents for quantities with the same base: When *Multiplying* terms with the same base, *Add* the exponents. When *Dividing* terms with the same base, *Subtract* the exponents. When raising a term with an exponent to another *Power,* then *Multiply* the exponents.

Mark
A button at the top of the screen during the GRE test. Clicking the Mark button will put a check mark next to that question on the Review screen.

Mathophobia
The quality of being frightened of math because you haven't done serious math in a long time and the only thing you remember is how bad you are at it. An imaginary affliction.

Mean
The average value of a set of numbers.

Median
The middle value in a set of numbers. Among an odd number of elements, the median is the middle number; among an even number of elements, the median is the average of the two middle numbers.

Mixed Fraction
A number that contains both an *integer* and a fraction, like $3\frac{1}{5}$.

Mode
The value that occurs most often in a set of numbers.

Multiple
The product of two *integers.*

N

Next
A button at the top of the screen during the GRE test. Clicking on the Next button will move you on to the next question. Simply clicking on an answer alone will not advance you to the next question; you must also hit Next.

Numerator
The top number of a fraction. If the numerator of a fraction equals zero (and the *denominator* does not), the fraction equals zero.

Numeric Entry
A question format on the GRE. Numeric entry questions have an empty box rather than answer choices. Some questions require fractions to be entered: These will have two empty boxes, one on top of another. Fractions don't need to be reduced to lowest terms (e.g. $\frac{1}{2}$, $\frac{16}{32}$, and $\frac{800}{1600}$ are all considered equivalent), and decimals do not need to be truncated (3.6 is the same as 3.600). Do not round decimals unless explicitly told to do so.

O

Obtuse Angle
An angle that measures between 90° and 180°.

Odd
Not divisible by 2.

P

π (pi)
The result when the *circumference* of a circle is divided by the *diameter* of that same circle. Roughly equal to 3.14159265358979323846 26…, but think of it as a smidgen larger than 3.

Parallel Lines
Lines within the same plane that will never intersect. On the coordinate axes, parallel lines have the same slope.

Parallelogram
A quadrilateral with two pairs of parallel sides.

PEMDAS
Acronym for Parentheses, Exponents, Multiply/Divide, Add/Subtract that describes the proper order of operations. Most calculators should perform PEMDAS automatically; however, it's important to check and make sure.

Perimeter
The sum of the lengths of all the sides of a polygon. The perimeter of a circle is called the "circumference."

Perfect Square
A number whose square root is an *integer*. The first five perfect squares are 1, 4, 9, 16, and 25.

Permutation
A possible arrangement of a certain number of terms when the order in which those items are selected does not matter. (See also *Combination*.)

Perpendicular
Intersecting at a right angle.

PITA (Plugging In the Answers)
Technique for answering multiple-choice questions without doing any algebra (see Chapter 4).

Plug In
A mathematical technique that changes an algebra problem into an arithmetic problem. Can be done whenever there are variables in the answers or the problem contains an unknown quantity that cannot be directly solved for. For more information see Chapter 4: Algebra, and How to Get Rid of It.

Prime Number
Any number whose only factors are itself and 1. The first twenty prime numbers are 2, 3, 5, 7, 11, 13, 17, 19, 23, 29, 31, 37, 41, 43, 47, 53, 59, 61, 67, and 71.

Product
The result of multiplication; the product of 12 and 4 is 12 × 4, or 48.

Pythagorean theorem
The formula you can use to find the length of the third side of a right angle ($a^2 + b^2 = c^2$, where a and b are the lengths of the legs and c is the length of the hypotenuse.

Pythagorean triplet
Any set of three integers that works in the Pythagorean theorem. The four most common Pythagorean triplets are 3:4:5, 5:12:15, 7:24:25, and 8:15:17.

Q

Quadrilateral
A polygon with four sides.

Quantitative Comparison
A question type on the GRE in which two quantities are given and you need to determine which is larger (or if the answer is impossible to determine). Often abbreviated in this book as "Quant Comp" or "QC."

Quotient
The result of division; the quotient of 12 and 4 is 12 ÷ 4, or 3.

R

Radical
Another name for a root; one might refer to $\sqrt{2}$ as "radical 2."

Radicand
The number inside the square root symbol.

Radius
The distance from the center of a circle to any point on the *circumference* of that circle.

Range
The difference between the greatest value and the least value in a set of numbers.

Rational Number
Any number that can be represented as the quotient of two integers.

Reciprocal
The result when the numerator and denominator are "flipped." The reciprocal of $\frac{2}{5}$ is $\frac{5}{2}$. The product of any number and its reciprocal is 1: $\frac{2}{5}\times\frac{5}{2}=\frac{10}{10}=1$.

Remainder
The result when a number does not divide evenly into another. When 7 is divided by 3, the remainder is 1.

Review
A button at the top of the screen during the GRE test. Clicking on the Review button will show a list of all questions, indicating which questions have been answered, remain unanswered, or have been marked to return to later. Make sure to click on Review before clicking on Exit Section, to make sure you have answered (or at least guessed on) every single question within a section.

Right Angle
An angle that measures 90°.

Right Triangle
A triangle in which one angle measures 90°.

Rhombus
A quadrilateral with four equal sides. (If all four angles are the same measure, then the rhombus is also a square.)

S

Scalene Triangle
A triangle in which no two sides have the same length and no two angles have the same measure.

Scratch Paper
Your saving grace on the GRE. Your favorite math buddy, more helpful than the calculator, more understanding than the computer screen. All work should be done on the scratch paper provided. In some testing centers, this may be normal paper, in others it may be a series of laminated boards. No matter what, the testing center will always provide scratch paper of some sort and necessary writing utensils. The first step in any GRE problem is to start setting the problem up on the scratch paper. Ask for more during every 5-minute break.

Square
A quadrilateral with four equal sides and four equal angles (each of which measures 90°).

Square Root ($\sqrt{\ }$)

The square root of x is the number that, when squared, results in x. For example, $\sqrt{16} = 4$, because $4^2 = 16$.

Slope

The rate at which a line is rising or falling within a coordinate plane. To find the slope of a line, use the formula $\frac{y_2 - y_1}{x_2 - x_1}$, which represents the "rise" over the "run."

Sum

The result of addition; the sum of 12 and 4 is 12 + 4, or 16.

Surface Area

The sum of the areas of each face of a three-dimensional figure.

T

Trapezoid

A quadrilateral with exactly one pair of parallel sides.

Trigger

A word or phrase within a GRE math problem that indicates exactly how you will answer that question. For instance, if a problem contains variables within the answers, then it is a Plug In question.

V

Variable

An element in an algebraic term or equation that is unknown or can vary. Variables are represented as letters.

Y

***y*-intercept**

The point where a line intersects with the y-axis. Represented by b in the equation $y = mx + b$.

Z

Zero

A number that has no value and therefore can't be used as a divisor on the GRE. Zero is not positive or negative, but it is even. It is also the additive identity, because any number plus zero equals that number ($m + 0 = m$).

ZONEF

An acronym to help you remember the weird numbers that can be helpful to plug in on Must Be or Quant Comp Plug In questions. Try normal, easy numbers first, but then try some of the ZONEF numbers on the remaining answers. ZONEF stands for Zero, One, Negative numbers, Extremely large or small numbers, and Fractions or decimals.

About the Author

Doug French has been a teacher, tutor, course developer, writer, and editor with The Princeton Review since 1992. He has taken dozens of standardized tests over the years and has never gotten a math question wrong. He is also the author of *Verbal Workout for the GMAT,* which should give you a sense of his awe–inspiring versatility.

Doug currently teaches calculus, precalculus, and geometry at The Hewitt School in New York City, where he lives with his wife and children.

NOTES

NOTES

NOTES

NOTES

NOTES

NOTES